# Synthesis Lectures on Engineering, Science, and Technology

The focus of this series is general topics, and applications about, and for, engineers and scientists on a wide array of applications, methods and advances. Most titles cover subjects such as professional development, education, and study skills, as well as basic introductory undergraduate material and other topics appropriate for a broader and less technical audience.

Ann-Perry Witmer

# Contextual Engineering

Translating User Voice Into Design

 Springer

Ann-Perry Witmer
Applied Research Institute
University of Illinois Urbana-Champaign
Champaign, IL, USA

ISSN 2690-0300             ISSN 2690-0327  (electronic)
Synthesis Lectures on Engineering, Science, and Technology
ISBN 978-3-031-07694-7      ISBN 978-3-031-07692-3  (eBook)
https://doi.org/10.1007/978-3-031-07692-3

This Springer imprint is published by the registered company Springer Nature Switzerland AG
The registered company address is: Gewerbestrasse 11, 6330 Cham, Switzerland

*To my sister, Sally Washabaugh, who has provided
a lifetime of encouragement, support,
unconditional love, and unfaltering confidence.*

# Preface

I write this Preface while sitting on a concrete slab outside the common house of a tiny indigenous community on the Altiplano of Bolivia, taking a break from talking with residents here about their infrastructure, their music, and their identity. I ask no questions of them about any of these facets of the community but instead engage with them on their terms, learning from them how to chew coca leaves, dancing with them, and exchanging information about how much a meal costs in the United States compared with Bolivia. Our team—including a research engineer, an engineering student, and two pre-eminent musicians (one Bolivian and one from the U.S.) arrived in this community barely four hours ago, and I already feel welcome, overcoming the wariness that so typically dominates an introductory meeting with the village to explain our presence. Our hosts are curious about us and our objectives in part because they've never been visited by people from the U.S. and in part because we are collaboratively researching how Western technology and music might both benefit by learning about indigenous traditions. The unusual focus of our work is a true ice-breaker for people, who are prepared to assume we're high-minded scholars preparing to explore unfathomable hypotheses in ways that exploit the populations we wish to understand.

As I listen to the braying donkeys and feel the breeze of a cool Andean summer afternoon, I'm transported back to an experience not so different than this of many years ago, when I first decided to see whether my engineering skills could be put to use in assisting a community in Guatemala. Back then, I entered the relationship with the village with typical Western hubris, predisposed to believing that my engineering skills would rescue a helpless, hopeless collection of people whose nation fell in the category of "low-income, developing." When I arrived in Guatemala, I found instead vibrant, creative, and hard-working indigenous people, who didn't want the newest technical designs—in fact, would likely reject them as excessive and unnecessary—and who innovated startlingly advanced engineering designs using whatever materials they had on hand. I started talking with them as a visiting friend, not as an engineer, and discovered I had much to learn from them. By the end of that trip, I ashamedly had to admit that my education from an elite engineering university could not match the applied designs they had created despite having only elementary school educations and no money at all. Over the course of the next decade, I visited more rural communities in more non-industrialized countries and learned more ways of thinking about technical design, first as a professional engineer, then as a student pursuing an unusual Ph.D. that merged engineering with the social sciences, and now as a research scientist and lecturer. Our research group studies how to gather and apply the knowledge that can be learned from other people and places to serve those people and places, and we're teaching the next generation of students to look and learn rather than dictate and develop. The discipline is now known as Contextual Engineering and has begun to branch beyond technical applications to encompass organizational dynamics, systems processes, and policymaking impacts under the sub-designation of Contextual Innovation and Practice.

Many engineering programs and engineering-based advocacy/aid agencies have adopted the concept of user-population engagement to improve engineering design, but the process typically used—often referred to as humanitarian engineering or engineering for society—still falls under the design process of the Western engineer rather than the pathway of the user community itself. It also fails to provide a framework that can teach the engineer about contextual conditions AND allow the engineer to apply those conditions directly to technical design decision-making.

Moreover, Contextual Engineering addresses the importance of the engineer's own mindset and predispositions in gathering information about the societal context and starts with self-evaluation for an engineering investigator. That self-evaluation leads the engineer to consider their own motivations, biases, and objectives, as well as the purposes and expectations of every other stakeholder in a project, most notably the user population itself.

How important is it to self-examine? Let me share a quick example as I watch the herds of cattle and goats return home from the fields down this sandy red road. I came to Bolivia to meet people and learn from them what they value, what they want, and what they are capable of doing, so that we can assist in establishing a clean energy–food–water technology training and research center for rural Bolivians. My musician colleagues seek to enhance their own understanding of music by learning and working with indigenous music traditions, which are deeply engrained in the Andean heritage. So for four hours today, we refrained from becoming defensive when asked what business we have been doing in this community, a question that could be perceived as unwelcoming or even threatening for someone unprepared to encounter skepticism. We shared information about ourselves with humility and empathy. We made friends rather than questioning subjects.

And here's what we've learned during those four hours—the residents of this community are curious about life in the U.S., but they deeply, deeply value the land on which they live. This is a society of creative problem solvers, as one Aymara woman demonstrated when offered some water from a friend; she had no cup so she simply removed her hat, poured water into it, then carried it, and drank from it. In a village home, we observed several basins used to carry and store water, all of which were made from auto tires that had been turned inside out, a circle of rubber from another tire stapled to the rim to create a sturdy flat vessel. These people have received government gifts for which they had no need, and the recreation field near where I now sit is totally vacant; for this reason, they look skeptically at anything they perceived to be associated with government "support." The Andean history runs deep in these people, evidenced not only by the way they dress but the way they arrange themselves in meetings, the way they request the floor when they want to share a thought, the way they wear their adorned herding whips on their backs as a combination of practicality and beauty. They value knowledge and have at least one person taking notes of everything they learn about us. Many of them carry small notebooks filled with words and sketches.

There is so much I haven't even begun to learn about my friends here and our team will spend another couple of weeks interacting with our hosts on the Altiplano to uncover a fraction of the knowledge that can help us but was completely hidden from us before. By shedding our expectations, humbling ourselves, and observing as much as we can, even in this short time we can already begin to formulate a contextual understanding of aspects that should govern the design of a training program for reliable water supply or improved agricultural production for these people in this place. While the application of my Western engineering training can improve the performance of their own innovations, it's the contextual conditions that will govern when to interject ideas and when to mimic practices already in place. It's a design process that is collaborative and engaging and respectful of our user community's capabilities. And that design process will occur in full partnership with our community, this small village at the top of the Bolivian Andes mountains.

This is Contextual Engineering. This is what I seek to share in the chapters ahead, both in terms of why we should do it, and how we can do it.

Champaign, IL, USA                                          Ann-Perry Witmer

# Acknowledgments

This manuscript is the result of the tireless research and critical thought of a talented collection of undergraduate and graduate students in the University of Illinois' Contextual Engineering Research Group. The dedication of these students in investigating so many facets of the relationship between technical design and societal context has provided enlightenment into the context that far exceeds my personal capabilities and understanding. Research Group members, individually and collectively, have advocated strongly for Contextual Engineering as an engineering-education discipline, and their passion and determination have contributed tremendously to legitimizing this avenue of research and application. Particular gratitude must be shown to the team that opened the door to investigating context for me: then-graduate students Keilin Jahnke and Eileen Walz introduced me to Dr. J. Bruce Eliott-Litchfield, who while serving as Assistant Dean in Engineering at the University of Illinois invited me to begin teaching my curious brand of engineering design. Further, I owe a debt of gratitude to Dr. Jennifer Bernhard, Director of the Illinois Applied Research Institute, for identifying a collaborative research value and creating the Contextual Innovation and Practice group within ARI.

Many of my students and colleagues have assisted me in wordsmithing and honing this Contextual Engineering text. Particular thanks go to Alexandra Timmons and Jessica Mingee (along with Jessica's parents) for combing through drafts while high in the Andes Mountains on field research and clarifying sometimes-complex issues.

Finally, I would be remiss if I failed to recognize the deep and abiding support that is provided to me daily by my husband, Stephen, and children Andrew, Grace, and Emma. The love and encouragement of family make everything possible.

# Contents

# The Basis for Contextual Engineering

**The Parable of the Street Dog**

Reina and Brava, named by U.S. visitors to the small Andean village of Calcha, Bolivia, are two stray puppies who have taken up residence in the visitors' residential compound. Both are skeletal, with ribs and hipbones protruding, and they hover warily around the main house of the compound, agilely ducking rocks or boots when they get too close to the kitchen. A scrap of food from a visitor's pocket one afternoon draws the dogs in, and within a few mornings, they're curled in tight balls outside the visitor's door. The visitor cautiously picks the shivering dogs up and warms them, surprised to find they instantly nestle into her arms and lick her with affection. It's a welcome comfort at a time the visitor is missing her own home and family.

Over the next 2 weeks, the visitor continues to warm the dogs, slip them bits of bread, and teach them to sit and shake, even bringing them into her room on one particularly cold night to burrow into her bed. Village residents are amused by the visitor's devotion to the dogs, but they allow her to indulge them as long as the dogs keep out of people's way and don't try to steal food that hasn't been offered to them.

For weeks after the visitor's departure, she thinks of the dogs, even going so far as to investigate the cost to fly them to the U.S. to live with her. Despite her love for the creatures, though, she never sees them again (Fig. 1.1).

© The Author(s), under exclusive license to Springer Nature Switzerland AG 2022
A.-P. Witmer, *Contextual Engineering*, Synthesis Lectures on Engineering, Science, and Technology, https://doi.org/10.1007/978-3-031-07694-3_1

**Fig. 1.1**  Reina reclines near her visitor-friend in Calcha, Bolivia

## 1.1    The Treacherous Path of Community Engineering in Unfamiliar Societies

One need not dive deeply into the academic literature to discover the passion with which the Western industrialized world strives to correct conditions of poverty and need through engineered infrastructure. Whether it is for safe drinking water or reliable transportation routes for food and supplies, the provision of infrastructure to non-industrialized communities by Western benefactors is a never-ending humanitarian effort. But case study after case study indicate that despite the bestowal of modern technology by highly trained technical experts, user populations continue to struggle in poverty and need, at best benefitting only temporarily from the relief they receive from the outside. Why?

Before we answer this question, let's look at a couple of engineering projects in rural Honduras.

The *barrios* (neighborhoods) of Llano Largo and San Antonio lie in southwest Honduras at the highest occupied elevation in Central America. For more than a decade, these two communities have shared a drinking-water system that supplies fresh water from a mountain spring to the valley below. The system was installed by a collection of international volunteer groups with no apparent local engineering input or collaboration, and it consists of a haphazard collection of pipes, unused reservoirs, and an abandoned well that was drilled in the valley before electricity was even available to power a pump. As a result, the water system's three *fontaneros* (system operators) spend their days opening and closing valves to control gravity flow from the mountain spring in Llano Largo down to San Antonio below. These *fontaneros*, hired jointly by the two communities, are technically adept and personally motivated to keep the water running for as many residents as possible, and the two communities have peacefully coexisted in sharing the water supply. A nearby Non-Government Organization (NGO) supported by U.S. interests also works closely with the community to provide ongoing support, resources, and technical training to the *fontaneros*.

But over time, both communities have added homes and additional water demands, taxing the system's already limited capabilities. The wealthiest residents have built new houses atop Llano Largo, above the spring elevation, while the poorest community members expand in the lowlands of San Antonio where the property is readily available at little cost. The president of the shared water committee, a Llano Largo resident who earned his wealth working in construction in the U.S., decided it was time to improve the water system so that all residents could obtain sufficient supply, regardless of where they lived.

A visiting Western engineer who examines the topography and water sources of the community is likely to rapidly propose a simple yet elegant technical solution. By tapping an additional spring source (Fig. 1.2) higher up the mountain for the wealthy homes atop Llano Largo, the existing spring could continue to serve the downslope portion of Llano Largo and the upslope portion of San Antonio. The dormant well could be retrofitted with a pump, now that the town is electrified, to serve the many homes of lower San Antonio.

**Fig. 1.2**  A community leader of the village of Llano Largo, Honduras, looks over a potential water source for a community supply

The three-part system could provide plentiful water to everyone at pressures neither too high nor too low to manage.

But this solution would have a catastrophic impact on the social fiber of the user population because the challenges of design for this system lay not in technical componentry but in the societal context. With the proposed solution:

- The poorest members of the community would draw water from a well, whose pump would generate ongoing power-utility expenses. Would the lowest income residents be the only system users required to pay those utility costs for their water, or should the expense be spread across the entire user population? If the latter option were chosen, would the wealthier homeowners atop the mountain agree to subsidize the poor residents using the well, even though they are from a different community altogether?
- Water quality testing for the new spring at the top of the mountain uncovers a number of contaminants that, while not acutely hazardous, produce a water supply that is inferior to the other two sources. Would the wealthy hilltop residents agree to limit themselves to this inferior water supply so that others could receive the superior water?
- How would the residents in the central part of the system feel about being isolated from their own *barrios* in favor of a new, hybrid neighborhood that combines portions of wealthy Llano Largo and poor San Antonio?
- Would the local NGO risk its reputation by trying to maintain the well pump—a technology it identifies as something with which it is neither familiar nor comfortable? Could the Western engineer overcome the NGO's discomfort by providing education and documentation on pump operation?

In contrast to Llano Largo/San Antonio's water system, neighboring Las Mesas was gifted with a modern water filtration plant that cost a large humanitarian NGO nearly $500,000 in materials and construction to build. While the NGO proudly displayed photos of the treatment plant on its website to boost its worldwide reputation for providing innovative water solutions to needy populations, Las Mesas residents have not used the treated water for several years. A brief visit there demonstrates why: while the plant's filter technology is modern (Fig. 1.3), no provisions were made for filter backwash and days after the plant was placed in service, the media became fouled and could not be cleaned. In the warm Central American air, the filters quickly began to incubate bacteria rather than removing them, and a month after start-up, community members reported suffering symptoms of water-borne illnesses like diarrhea and intestinal pain. Las Mesas system *fontaneros* reported that the plant's designers had spent no more than two days on-site, and they provided the barest operational documents and training once the facility was completed. The *fontaneros* invented their own filter-cleaning method, periodically closing the outlet valve to flood the media so they could rake off any algae that floated to the surface. This provided no regeneration for the bottom layers of the filter, which continued to produce harmful bacteria in the tropic heat.

**Fig. 1.3**  A filter vessel in the Las Mesas water treatment plant, Honduras, which is cleaned by flooding the tank and sweeping off floating algae with a rake

These stories demonstrate multiple issues that confound international engineering efforts and provide an explanation for why these efforts so often are ineffective at addressing local needs:

1. While designer and user may have the same root objective of supplying safe drinking water for a user population, the motivations that drive the engineer's selection of technology design may ignore the needs and conditions of the user, resulting in infrastructure failure or rejection;
2. Technological design that fails to incorporate the conditions of people, place, and time often will fail, be abandoned, or even fall victim to sabotage by the users for whom it's intended. Worse yet, a contextually uninformed technology may tear at the social structure of the user population and create rifts along lines of cultural identity, economic class, or social status;
3. One may assume that education and training will overcome the complexity of the design. But this assumption not only is naïve, but also can be insulting, patronizing, and imperialistic because it neglects to incorporate local knowledge and local capabilities, not to mention local values and beliefs associated with the technology.

These examples demonstrate that a technical design must be interwoven with sociocultural identity if it is to achieve the objective of addressing a user population's physical needs effectively and sustainably. This is particularly important for rural communities where identity is integrally tied to location, where indigenous knowledge is valued and effectively applied, and where exposure to globalized technology is limited and sometimes discouraged.

To better understand how an engineer's genuine desire to help others can result in the exact opposite outcome by not considering context, we need to spend a little time looking more closely at why—and whether—Western technology can or should address the needs of the non-industrialized world. To do this, we must step away from our technical training and engage in sociological inquiry to better understand the dynamics associated with industrialized-world engineers providing services in non-industrialized locales. First, we look at the conditions that have encouraged engineering designers to undertake international infrastructure projects in rapidly growing numbers during the past 50 years. Second, we seek to understand how those projects are evaluated—and by whom—to determine whether they may be regarded as meeting the specific objectives of all the participating stakeholders, from user to designer to funder to manager. And third, we must consider the particular significance of rurality in the application of engineering technology because rural populations harbor the most well-preserved traditions of identity and values. We attribute this preservation to the rural population's isolation from the globally connected, indigenously diluted, cosmopolitan world-population paradigm.

## 1.2    The State of Technical Design in the Globalized World

As the nations of the world progress in embracing global communication and establishing global interdependence, agencies and organizations increasingly take responsibility for providing engineering services to societies that differ from their own, particularly in terms of need (Nieusma & Riley, 2010).

The planet's technical experts grow more confident daily in their ability to gain access to societies previously considered "untamed," both in terms of remote communication and travel. Whether by visiting or by Googling, they can witness the disparities between industrialized-world sophistication and non-industrialized privation and seek to advance the non-industrialized populations' access to modern technology. But access doesn't translate into a clear understanding of the societal differences that may or may not make a particular technology appropriate or desirable to meet the recipient society's needs (Ika & Hodgson, 2014). In fact, a variety of non-technology influences intermingle within each user community to generate a particular response to Western technology that can appear not only unique but also seemingly unpredictable for a given engineering approach (Hjorth & Bagheri, 2006). While the interrelationship of these influences varies widely from one non-industrialized population to the next, engineers continue to seek the "silver bullet" solution to global infrastructure needs, oblivious to the sociological perspective that a universal solution does not and cannot exist in a globalized world (Haraway, 1991).

Local characteristics often provide protection against power subversion, and the implementation of a broad Western solution can result in diluting or overpowering that local protection.

In our Contextual Engineering courses at the University of Illinois, we challenge students to consider whether engineering support from the Western world to non-industrial populations not only misses the mark as support, but also actually is a thinly veiled modern form of colonization. Addressing this possibility typically generates considerable discomfort among students in the classroom. "But we're *helping* them to have a better life!" students are prone to exclaiming. The question we always ask back is, "By whose definition?"

In one particular classroom exchange, I challenged a student to explain why a rural Central American community should compel its residents to chlorinate their water, even though they feared that the chlorine might somehow infiltrate their coffee farms and ruin the value of their crop.

"They have to chlorinate!" he exclaimed. "We're giving them the opportunity to extend their lives by drinking safer water!"

"Ah, so it's about extending their lives?" I asked.

"Of course!" he responded. "Everyone wants to live a long life."

In ancient times, the Mayan ancestors of this community held athletic tournaments in stadiums that now lie in ruins, I told him. During those tournaments, the winning team celebrated their victory by being put to death, allowing them to achieve their eternal glory more quickly. The student struggled to respond.

"Let's think about this in a more modern way. If you're elderly and you live in a place where healthcare is very hard to get to and may not be reliable, do you want to spend your geriatric years burdening your children, relying on them to feed you, bathe you, ease your suffering? They have their own children to feed and care for. Would that make you want to live a long life?"

The student simply shook his head, acknowledging that he assumed everyone would value a long life, just as he did.

While I was not actually condemning the practice of chlorination during this discussion, it was important to guide the student toward recognizing that his insistence on using the "everybody knows…" argument may actually lead him into conflict with a user population's knowledge, values, or capabilities. If one relies on her understanding of what a population should believe, rather than learning what they do believe—and equally importantly, why—she takes from that population its power of self-determination.

And when self-determination is gone, how can she expect the population to act independently to maintain the infrastructure placed in their care?

This misplaced notion that all people share the same beliefs and needs, which are articulately expressed by Western societies, is by no means a recent circumstance, nor is it accidental. In reviewing the history of international infrastructure interventions and drivers that lead engineering designers in the industrialized world to address physical needs elsewhere using their own expertise, the emergent condition that coincides with international service and responsibility is the Globalization Project, which took hold in the 1970s and redefined space and place in a globalized network (Woods, 2007). A very basic definition of globalization is provided in What Is Globalization? (2016):

> Globalization is a process of interaction and integration among the people, companies, and governments of different nations, a process driven by international trade and investment and aided by information technology. This process has effects on the environment, on culture, on political systems, on economic development and prosperity, and on human physical well-being in societies around the world.

The impact of globalization is particularly notable in the "global countryside," which has become a surrogate for "the local" amid "the global," creating a hybridization of both. The result of this hybridization can be either positive or negative, depending upon not only human-based conditions but ecological, topographical, and other non-human conditions as well (Massey & Massey, 2005). Sociologists assert that the dynamics associated with global countryside hybridization rely not only on economics but on social, cultural, and political processes as well, all of which can come together in unique combinations to result in a net positive or negative experience for the rural society under investigation. But while the hybridization of these processes is apparent, it is important to note that globalization as a whole is driven "by international trade and investment," or more bluntly, money.

If one has doubts about this, please consider one of the largest and most infamous failed efforts to support development through infrastructure (Chambers, 2009). The Play-Pump initiative began in South Africa, intended to leverage children's playtime activity into water provision by using a playground merry-go-round to drive a pump that could draw groundwater supply into a tank plastered with advertisements, the sales of which could generate income to maintain the water infrastructure. The entire contrivance both captures and leverages the Western notions of marketing, monetizing, economizing, and exploiting for the purpose of social good. Western organizations ranging from the U.S. Government to the World Bank to celebrity Jay-Z donated tens of millions of dollars to support what they considered a transformational technology, but savvy organizations challenged the functionality and purpose of such a device. Beyond the notion of commercializing rural non-Western communities with billboards on water tanks, the pumping process itself ignored a variety of contextual issues. Children in many parts of the world take on household duties like cleaning, caring for younger siblings, working in fields, or cooking from the moment they can walk, and they lack the boundless leisure time in which Western children indulge. Even if they did desire to play on the PlayPump merry-go-round, by one calculation they would have to play non-stop for 27 h each day to produce a minimum daily water need of 15 L per person to a village of 200 people. And yet, this solution combined the Western experience of carefree childhood days and a thirst for opening new markets through advertising, making the PlayPump a perceived win–win for development support.

Venugopal (2018) provides an argument that the PlayPump program's architects could have used to justify their effort as a success. He suggests that the valuation of a development project is wholeheartedly dependent upon the perspective of the powerful at the cost of the powerless, an overriding theme of globalization:

> . . . The fundamental impulse of the dominant party is to extract benefit from the subservient one. Development is thus the pursuit of those self-interested objectives, while the language and rhetoric of upliftment and selfless generosity euphemize that reality and provide it with the legitimacy to render it acceptable.

While the PlayPump example makes Western forces look almost predatory in providing infrastructure assistance to non-industrialized societies, Lacy (2009) further explores the relationship between globalization and the rural world by examining the impact of globalization-driven desires to "empower" rural communities through science and local food system innovations. The ways in which a globalized knowledge of science and technology is disseminated to rural societies around the world has become an essential factor in modifying those societies' self-identity and self-determination. However, Lacy argues, globalization has shifted the definition of a community's worth from a social unit to a commodity-production site, which now places an imperative on rural communities to become contributors to the global economy rather than to remain self-contained pockets of place-defined cultural curation. Many of these societies resist globalization, though, much as they resisted colonization by the European empires that imposed new definitions of civilization upon them in the past. Virtue Theory provides justification for this resistance to community "development" and loss of local authority. The three facets of Virtue Theory Lacy focuses on are how: (1) commodification of goods and services may undermine virtue, (2) development may make the performance of tasks routine and unreflective, and (3) development may interrupt the continuation of family traditions, which are associated with a virtuous lifestyle. Violations of these virtues can significantly influence a community's governance structure, turning citizens from active participants and vibrant members of a social unit to producers, voters, and consumers who are passively involved in the unit, disempowered from supporting the inherent identity of that unit. The weakening of local responsibility and initiative, particularly through the promotion of globalized science, technology, and education, is particularly troubling in this analysis, because it points directly to a potentially dangerous impact associated with outside technical designers relying on their globalized expertise to support a locally defined and governed rural society:

> The products of science are contextually specific constructs that can be understood only with detailed knowledge of the social conditions of their production. These include decisions about

the choice of problem, what resources to allocate to the problem, how to conduct the research, what to consider as results, and how to interpret the findings (Lacy, 2009).

Lacy is not the only scholar to recognize the significance of locally defined products of science. Bruno Wambi (1988), the former president of the Congolese Association for the Development of Library and Archival Documents, once argued that "technology is like genetic material—it is encoded with the characteristics of the society which developed it, and it tries to reproduce that society."

In consideration of whether indigeneity becomes less relevant when connections are formed in a globalized context, Ferguson (2012) proposed that identity, and thus the "genetic material" of a society, can be lost through globalization. Ferguson considered how the people of Zambia abandoned their cultural identity during colonization, adopting instead the perceived superior deportment of the Anglophone colonizers. In the post-colonial period, however, Zambia lost its relevance to the industrialized world and not only was devalued by its colonial sovereigns, Ferguson stated, but also was left to feel foolish at having believed itself a peer of the Europeans in the first place. If technical designers are tempted to consider engineered infrastructure as a device of connectivity between technology designers and users, Ferguson's cautionary tale of Zambia provides compelling concern for the impact that such connectivity may create upon the indigenous identity.

Alverson (1977), meanwhile, examined the disconnect between indigenous societies and volunteers from the Global North who seek to support them, as well as the resulting relational dysfunctions that prevent progress. Describing Botswanan practices that ranged from showing up late for meetings to standing very close to others, Alverson recounts how Peace Corps volunteers ascribed nefarious motivations to these actions and took offense. Alverson demonstrates that while U.S. volunteers witness behaviors that appear familiar and meaningful, they lack the indigenous understanding to interpret those behaviors properly. From disparities in the consciousness of time to understandings of the value of candor, volunteers failed to translate behaviors based on indigenous identity rather than their own, leading to consistent failures to make progress in identified global-humanitarian collaborations.

A few years ago, at a conference on technology and development, I shared the experience of working with a Honduran community that was divided into several sectors, one of which was on the outs with the others. When we arrived in the village with our NGO partners, the outsider sector immediately came to us and told us of their particular water needs. This sector lay at a higher elevation than the community's water source and so could not be provided with the supply by gravity. Without electricity, the community could not pump water to this sector, and the creation of a power source and pumped system would generate considerable ongoing expense for system operation and maintenance. The community *junta* (an elected board that oversees water operations) considered its options before determining that it would not burden the entire community (Fig. 1.4) to

**Fig. 1.4** A father and child from a divided rural Honduras community participate in water-sanitation exercise with NGO partner

serve a small portion of the population at great expense, and it elected to move forward with a system that only provided water to homes that lay below the source elevation. After I described this project to the conference, one gentleman who had considerable experience working with NGOs in East Africa angrily demanded to know why I didn't insist on serving the entire community.

"It was the community's call, not mine," I responded.

"You should have made the community provide water to everyone," he responded, raising his voice. "If they wouldn't serve the upper sector, you should have told them you wouldn't build them a water system at all."

"Let's think about this from their perspective," I responded. "The community and the upper sector haven't gotten along or worked together for a very long time. What do you think would have happened if I had insisted that they work together here? Would the entire village have agreed to pay for the pump operation? Not once I left town. And what right have I to say that I would deny water to the entire community unless it agreed to my terms?"

Ultimately, my role in working with this community was to provide an infrastructure design, not to heal old rifts or apply my own values to a population that functioned very differently from my own. Few engineers are trained to modify the social structure of a population, yet the notion that working with a population that is non-industrialized and non-globalized is equivalent to working with populations in need of being taught how to function is a pervasive driver for many engineers working outside their own lands. In fact, development engineering, as international design for non-industrialized users often is called, suggests that the objective of an engineered project is to improve the condition of the society in terms of economics, health, and political-conflict abatement (Lentfer, 2017). These are heavy responsibilities to be placed on an engineer charged with creating a civil infrastructure design, particularly outside of their own culture of experience, raising the question of whether an engineering project should even be considered a form of "development" (Witmer, 2017). Nieusma and Riley (2010) consider engineering and development to be intertwined, and they proposed that an engineer's focus on technology may neglect the considerations of economic and cultural structures that direct development interventions. As a result, they conclude, technological interventions may be functional but social-justice conditions may be placed at risk. Technical functions "tend to occlude social power imbalances and epistemological divergence, leading to projects that inadvertently extend social injustices." Such assertions are valid, as long as they aren't used to justify development actions in the name of engineering design.

Recognition by international service engineers that user communities may have different values, behaviors, and standards, then, becomes critical to creating an accepted and understood technical infrastructure. This assumes the technical design is intended to solve a user community's physical need rather than to extend the global engineering paradigm to "outsider" rural societies, as Ferguson suggests. The ability of the engineer to focus only on local standards, as Alverson indicates, may be difficult to achieve without a mindful awareness of societal disparities between user and designer.

You may have noticed that many of the scholars cited in this section have considered for decades the detrimental impact of Western technology standards upon non-industrialized populations. These are not new concepts, and yet they continue to be ignored by engineers under the assumption that the science and math of technical design somehow transcend the complications of working with people, identities, and values. Contextual Engineering challenges this assumption.

## 1.3    Infrastructure Interventions by the Industrialized World

Armed with a sociological understanding of the drivers that initiate a relationship between an industrialized provider and a non-industrialized recipient, along with acknowledgment of differing perspectives that can inform a stakeholder's perception of project effectiveness, we can now look with a new eye at how the engineering scholars have regarded "international development projects."

In fairness, the engineering literature has sporadically explored the influence of local social characteristics upon the durability and sustainability of an engineered infrastructure, but these explorations often consist of case studies for a specific locale, assuming that a lesson learned in south Asia may be applied with equal effectiveness to a community in Andean South America or sub-Saharan Africa. The reason for this assumption may lie in the recognition that past engineering efforts to export Westernized technological solutions to non-industrialized societies have frequently been identified as failures (Bouabid & Louis, 2015). Looking specifically at efforts to provide safe, reliable drinking water to non-industrialized countries throughout the world, estimates of infrastructure failure for outside-delivered interventions range from 30 to 60%, and at least one NGO predicted that 50% of water systems constructed by non-Hondurans in Honduras would fail within five years (Stottlemyer, 2017).

We understand that global drivers, motivators, and forces have brought Western engineers to seek solutions for non-industrialized user populations. Now, we must consider the mechanisms by which parties regard the value of this relationship. Implicit in this statement is the assumption that all parties to the relationship share a clear definition of "success" versus "failure." In reality, this could not be farther from the truth, and the sociological literature demonstrates that the concepts of success and failure are more dependent upon the perspectives and personal motivations of each stakeholder than upon actual infrastructure performance.

Dunn's (2017) look at the Polish pork market after the dissolution of the Soviet bloc provides strong evidence that regulations and policies of the industrialized world not only may disagree with local values but in fact may even adversely affect indigenous self-governance, economy, and identity. In her examination of the pork market, Dunn found that Polish people prefer a fattier pork, though it's considered inferior to European standards. Additionally, the cleanliness standards for European meat production required significant equipment investments that drove Polish butchers out of business. The combination of meeting Western meat standards and violating local preferences led to a black market for Polish pork and the loss of jobs for many in the production industry. The imposition of a "global standard" for pork production upon a previously disconnected society, then, not only undermined its ability to address its own needs but excluded it from a greater pork market by labeling its product as inferior. Dunn demonstrates that the Western standards foisted upon a non-industrial society conflicted with the unique practices, tastes, and capabilities of that society and ultimately eliminated it from participating in globalized markets. The same thing can occur with the imposition of Western engineering standards.

While visiting a rural Guatemalan community and meeting with its water committee, I witnessed an example of this. As we walked along the path of the gravity-flow pipeline that eventually reached the town from a spring above, the committee members showed me a spot where the pipe route rose over a small hill.

"Do you ever have issues with airlocks in the pipe, when the spring is not flowing or demand exceeds supply?" I asked them.

"Of course we do!" one of the men said, while kicking aside a rock that covered the buried pipe. Sticking out of the top of the PVC was a carefully whittled stick, which he plucked to release a stream of water from the pipe hole. "When the pipeline doesn't flow, we just pull this out, and when the water flows again and everything works well, we put it back in its place."

Of course, design standards for Western water systems do not allow for whittled sticks to act as air release valves in public water supplies. Some of my Western colleagues would advocate for replacing the hole and stick with a manufactured air-release valve, which can be notoriously difficult to maintain in systems such as this because of clogging, sticking, and deteriorating seals. If I were to insist the community replace the stick-and-hole device with a valve, of course, they would take pride in the advanced technology until it failed. Then they would disable or remove it, carve another stick, and return to a technology they understood and innovated.

From the perspective of Western standards adherents, then, the acceptance of stick-and-hole air release valves—much like the development of a Polish pork black market to circumvent European Union quality standards—could be viewed as a failure. From the perspective of the users, though, it may be viewed as a success because it served local needs in a cost-effective, easily adopted way that could function indefinitely with little care. From the perspective of those upon whom the standards were imposed, Dunn demonstrates, globalized regulations only serve to control and constrain conditions that may eliminate populations from the globalized marketplace.

The notion of imposing regulations or standards that conflict with local needs and sensibilities is not limited to international relationships. Grigg's (2017) review of the Flint, Michigan, drinking-water supply crisis examines an intra-industrialized failure to address user needs for drinking-water safety, attributing that failure to a series of non-technical conditions that led to the abdication of responsibility. In that crisis, a predominately low-income population's water supply was changed without accounting for concurrent changes in water chemistry. The result was a corrosive water supply that rapidly leached toxic levels of lead from household plumbing into drinking water—a clear violation of federal health protections required of public water supplies—and a failure by regulatory agencies and water-utility managers to accept responsibility or address complaints. Drinking-water engineers consider policies and standards of the U.S. Environmental Protection Agency (USEPA) as being among the most effective in the world, and these standards often are used as measuring sticks for projects conceived and implemented for non-industrialized populations. Griggs notes, however, that the regulatory framework failed Flint customers because of a combination of community economic distress, political disregard for customer well-being, and general ignorance of the linkages within a regulatory framework that can influence behaviors and outcomes. The failures that led to the Flint water crisis, when firmly placed within the framework of an effective regulatory structure, can be viewed as a deterrent to the imposition of regulations and policies across societies and cultures.

There is a growing body of scholarship that acknowledges the importance of performance in international infrastructure design, but it typically is coupled with a recognition that many functional projects still fail to meet the standards of the designers themselves because of sociocultural conditions (Starkl et al., 2013). Scholars often use case studies to demonstrate project performance in the context of user acceptance, but they rarely reflect on the interaction of the technical designers' own context in design decision-making and implementation. Let's look at some of the trends demonstrated as well as the assumptions that scholars make in assessing project outcomes.

We've already introduced the variability of understanding in the terms "success" and "failure." Now let's look at what they mean to the Western engineer. Starkl et al. (2013) provide one of the broadest evaluations of engineering effectiveness in non-industrialized community water systems, espousing the premise that a general quality standard may be applied to systems throughout the world. The authors examined 60 projects in India, Mexico, and South Africa to see how many were regarded as "apparent successes" by local experts compared with whether they were "actual successes" in the view of international expert teams, and how both these statuses might be compared with "social successes" as determined by consumer use. The paper's authors define success as a water infrastructure that delivers the expected services without unacceptable risk. Of the 60 systems examined, 47 were "socially successful" in that they were used. Local experts considered 29 systems to be "apparently successful" with another 7 partially apparently successful. But when global experts were consulted, only 13 systems were determined to be "actually successful" because they adhered to the experts' safe drinking water standards. This means that global experts considered 47 water systems to be at least partial failures, while users themselves considered only 18 systems as failing to fully meet the needs of the user population. Starkl's analysis of the data considered whether policymakers must have a stronger hand in maintaining such systems, advocating that "At least in economically advanced developing countries...it is necessary to enforce standards and procedures that are comparable with those of industrialized countries." One may recognize the similarity of the conclusion to the description by Dunn of how imposed standards of measurement and evaluation drove Polish pork producers out of business or underground.

The adherence to industrialized-world standards and policies was further advocated by Shannon et al. (2008). As with Starkl, Shannon discounted user opinion of success in determining what the authors viewed as a global standard for intensively disinfected and treated water supply, regardless of access, indigenous belief, or economic means. Shannon described the ubiquitous world problem of inadequate access to fully disinfected water supply, neglecting to acknowledge the even greater problem of inadequate access to consumable water. Absurdly, Shannon acknowledges that ultraviolet disinfection may lack the ability to inactivate viruses in supply while ignoring the pervasive lack of reliable electricity in non-industrialized societies, which is necessary to power an ultraviolet treatment process. Yes, there is a need to decontaminate water supply containing heavy metals, nitrates, or aromatics. But imposing a requirement to remove contaminants per Western standards at a level that couldn't even be detected 30 years ago sets a Dunn-like showdown of control versus accountability. Scientists since the early 2000s have been monitoring minute levels of artificial sweeteners in U.S. public water supplies such as Lake Mead (Mawhinney et al., 2011) but it would be absurd to judge the success of a water system in West Africa today upon whether it can reliably confirm the absence of sucralose. The expert focus on technology to assure the safety of the world's water supply creates the potential for making that water supply inaccessible to many people, potentially putting access further out of reach if policymakers focus on global quality standards before local availability.

From these examples, one could conclude that engineering effectiveness is the ultimate objective of infrastructure interventions, yet engineering effectiveness has been defined in multiple ways, most typically from the viewpoint of the Western "experts" who design and implement infrastructure interventions for societies that differ in cultural identity, values, and experience. For drinking-water supplies, for example, the metrics of effectiveness are broad and vague, and include the Millennium Development Goals set by the United Nations and the 1992 Dublin-Rio Principles of the UN Committee on Economic, Social and Cultural Rights (Amin et al., 2015). Carter et al. (1999) defined effectiveness—in the form of impact objectives—for a water system in terms of minimum daily per capita volumes, minimum times spent hauling water, improvements to water transport technology, minimal downtimes for water delivery, equity in service provision, and decreased contamination.

In contrast to this generalized (and very difficult to define) evaluation of success, other authors focused on specific standards to which industrialized societies adhere, such as the Safe Drinking Water Act in the United States. Shannon et al. (2008) invested great effort in discussing the development of water-purification technologies to support the "developing[1] world," indifferent to the notion that many societies refuse to disinfect supply because it conflicts with their spiritual beliefs, and many more societies lack the resources to implement and maintain ultraviolet disinfection technology so that they can avoid disinfection byproducts from chlorination (a contaminant unregulated in the United States until 1999, when detection capabilities became sufficient to identify the constituent).

---

[1] The term "developing" is not used in this text without quotation marks because it implies that one society has not progressed along its evolutionary path until it achieves the conditions dictated by those societies who are considered "developed." Except when quoting others' material, we will refrain from using comparative language and instead defer to the terms Western and non-industrialized to represent societies.

But what if we consider project success as a measure of user population adoption, sustainability, and satisfaction instead? The literature is mostly silent on identifying successful projects directly from the perspective of the user, and more than a decade of personal experience working with non-industrialized populations reaffirms that no one ever asks them the very simple question, "What would make this project a success for you?".

We've discussed here the evaluation of project success, at least from the perspective of the technical designer. Now let's turn our attention to examples of infrastructure failures in the literature, where recognition of sociocultural considerations in engineering design becomes a bit more fashionable, even if those sociocultural conditions are transient from location to location. Amin et al. (2015) evaluated the implementation of water supplies in India by an NGO and an Australian educational institution to conclude that best practices for system implementation rely on awareness of communal wealth and the installation of household taps. Using household surveys for 10 case-study communities to assess the effectiveness of service, the authors developed analyses correlating median income to the acceptability of service. A glaring omission for this paper, though, was that the authors inferred causality from correlation, which has the potential to drive readers toward the assumption that "best practice" is to install water systems in the highest income communities to ensure success.

By contrast, Harvey and Reed (2006) considered community-managed water supplies in sub-Saharan Africa and the distinction between community participation and community management to ensure water system sustainability. The methodology of investigation for this case study of African systems is unclear, as is the very definition of what constitutes an "adequate" water supply provided by a community system. However, the conclusions of the paper were that African communities are capable of participating in system development but should not be expected to manage the system. Instead, the authors concluded that "appropriate institutional support" is needed to ensure system sustainability. In the alternative, they added as an afterthought in the last paragraph of the paper: "If user communities are to be truly empowered and granted true decision-making authority, they should be given comprehensive information needed to make informed decisions, without being pressured to follow the preferences of the facilitator... Unless such an approach is taken, use of the term 'community development' in relation to rural water supply will remain rhetoric rather than reality." Two concerns arise from this paper: First, there is an implicit assumption of conformity among communities of the many nations of sub-Saharan Africa. Differences in social relationships, cultural mores, and societal values are utterly neglected in the broad-brush depiction of the failure that community management will bring to a water system. Second, one cannot articulate based on reading the article exactly what that failure would look like, other than to state that there would be an inadequacy of performance associated with community-managed systems. The lack of transparency in assessing not only sociological conditions but also very basic engineering considerations associated with the subject communities is troubling.

In contrast, Vaccari et al. (2017) provided significant detail and analysis in evaluating the appropriateness of cooking technology for application in the Logone Valley on the border between Chad and Cameroon. The authors meticulously identified both technological and socioeconomic drivers toward community residents' choice of cooking technology, and they employed a series of criteria to make their analysis more robust. This case study distinguishes itself by acknowledging that not all societies are monetized, complicating economic analysis, as well as exploring user practices to resist the trap of characterizing certain technologies as "gender equitable" when such equity may conflict with the female user's desires and values.

Multiple additional case studies were reviewed as part of this process, but the findings of the literature are consistent and may be summarized thusly:

- There is a varying level of sophistication in analyzing the impact of infrastructure intervention.
- There is a predisposition toward using a case-study experience to generalize best practices across cultural lines.
- There is no clear definition of what criteria determine an intervention as successful or failed from the perspective of the user, since the perspective of the author (often also the technical designer) dominates the analysis.
- The opinion of the user with regard to project performance is rarely considered, and when it is, it typically provides a demonstration of the contrast between user perception and Western performance standards.

## 1.4  Why the Rural Matters

While the place of rural society has been briefly discussed above in relation to the Globalization Project, a more specific exploration is necessary to understand rural society as an agricultural center, a custodian of place-based tradition in a globally networked world, and a center for environmental preservation—and how that identity conflicts economically and educationally with the urban.

Throughout the globe, rural societies act as the kitchen pantry of the world's local identities, stocked with a diversity of knowledge, beliefs, and value systems (Creed & Ching, 1997). Indigenous beliefs, practices, and knowledges in rural communities throughout the world are as diverse as the flora that has evolved in different regions based on climate, topography, and geology.

In the twenty-first century, though, rural communities also are home to the world's poorest populations. As Lipton (1977) observes in "Why Poor People Stay Poor", "the worst-off one third of mankind comprises the (rural) underclass of the Third World." Development policy-makers cast their eyes toward rural societies as targets for poverty reduction, but trends indicate many peoples are self-managing poverty by migrating to population centers (Imai et al., 2017). Urban melting pots of global bombardment, with a blur of fast-food restaurants, product advertisements, and access to viral internet memes, rapidly dissipate the place-based knowledge and identity of local societies (Creed & Ching, 1997). Drivers of this migration to urban areas frequently are economic, but a significant reason given by migrants is a desire to live in conditions that more adequately address basic human needs—safe drinking water, improved sanitation, electrical power, and passable roadways among them (Radhakrishnan & Arunachalam, 2017). Compare, for example, images from rural Calcha (Fig. 1.5) and urban La Paz, Bolivia (Fig. 1.6), for a clearer understanding of the cultural and economic divide between the two types of population centers.

International aid organizations have provided engineered infrastructure interventions to non-industrialized societies for decades, dating back formally to the time of the Bretton Woods Agreement of 1944. But as the business of development has grown with the onset of the Globalization Project in the 1970s, the emphasis of these interventions has been on reducing the volume of people living without basic services (Krause, 2014). Initiatives such as the United Nations Millennium Development Goals of 2015 effectively set quotas for the reduction of the number of people without access to clean water or improved

**Fig. 1.5**  The central square of rural Calcha, Bolivia, largely abandoned

**Fig. 1.6** Cosmopolitan La Paz, Bolivia, seen from a *teleferico* car high above the downtown

sanitation, driving organizations toward supporting urban interventions to optimize service delivery. As a result, the largest and most-resourced aid agencies have placed a strong focus on population-center interventions, providing generalized and "scalable" guidance for use by more rurally focused organizations but leaving the global countryside largely in the hands of less well-endowed volunteer groups, missions, and NGO interventions (Matthew et al., 2016). Rural infrastructure support, then, often falls to organizations that are less equipped and less resourced to address local needs while relying upon global attitudes and values that conflict with local identities. The remnants of failed infrastructure systems that litter the global countryside bear witness to the widening gap between rural and urban intervention effectiveness (Witmer, 2018).

Further validating the support for urban centers are the assumptions that cities are more fertile, more esoteric, and more intellectual than rural societies. As discussed in the introduction to their collection of essays, Creed and Ching (1997) directly confront the difference in power and respect bestowed upon the "urbane" city dwellers of the world and the "rustic" rural societies, who are regarded as lacking sufficient intellectual value to the extent that even academic researchers neglect to consider their contributions. This attitude was demonstrated in discussion after a paper presentation by Mincyte (2018), who stated that there is an "urban aesthetical view of the landscape" that is in conflict with rural sensibilities and beliefs toward land. In fact, she said, many urbanites regard rurality as a leisure-time place to experience nature rather than a center for livelihoods and a location upon which a nation's food security is dependent. Assuming one values the preservation of rural knowledge, values, and economic activities that reside in the countryside, in much the same way one values the preservation of an endangered species, it may be necessary to rethink the worth of rural societies and the importance of addressing rural interventions in a new way that supports specifically place-based needs. This section will explore the conditions that drive rural residents into forced urban migration, along with the state of infrastructure interventions that contribute to destabilization of rural community populations when they don't contextually address rural needs.

Data from the United Nations Department of Economic and Social Affairs (2018) portray a strong trend toward urbanization and decay of rurality throughout the globe. Figure 1.7 depicts past and projected trends in rural and urban populations for the entire world (in red) and particularly for low-income countries (in blue). The very definitions of urban and rural are somewhat challenging to pin down, since each nation chooses its own definition for rurality. The U.S. Census defines rural communities as those with less than 2,500 inhabitants, while Iceland's definition limits rural communities to those with less than 200 inhabitants (Timmons, 2021). Though most low-income countries have been more heavily populated in rural areas than urban since the start of modern development efforts, the urban population is projected to overtake rurality in this sector before 2050. Globally, urbanization overtook the rural population in the late 2000s, and Fig. 1.7 graphically demonstrates that future population growth is projected to dominate in cities, while rural populations are projected to decline despite an increase in overall population expansion. Some of this transition to urban dominance may be attributed to the migration

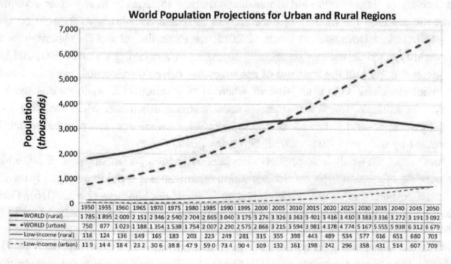

**World Population Projections for Urban and Rural Regions**

Population (thousands)

| | 1950 | 1955 | 1960 | 1965 | 1970 | 1975 | 1980 | 1985 | 1990 | 1995 | 2000 | 2005 | 2010 | 2015 | 2020 | 2025 | 2030 | 2035 | 2040 | 2045 | 2050 |
|---|---|---|---|---|---|---|---|---|---|---|---|---|---|---|---|---|---|---|---|---|---|
| WORLD (rural) | 1 785 | 1 895 | 2 009 | 2 151 | 2 346 | 2 540 | 2 704 | 2 865 | 3 040 | 3 175 | 3 276 | 3 326 | 3 363 | 3 401 | 3 416 | 3 410 | 3 383 | 3 336 | 3 272 | 3 191 | 3 092 |
| WORLD (urban) | 750 | 877 | 1 023 | 1 188 | 1 354 | 1 538 | 1 754 | 2 007 | 2 290 | 2 575 | 2 868 | 3 215 | 3 594 | 3 981 | 4 378 | 4 774 | 5 167 | 5 555 | 5 938 | 6 312 | 6 679 |
| Low-income (rural) | 116 | 124 | 136 | 149 | 165 | 183 | 203 | 223 | 249 | 281 | 315 | 355 | 398 | 443 | 489 | 534 | 577 | 616 | 651 | 680 | 703 |
| Low-income (urban) | 119 | 144 | 184 | 232 | 306 | 388 | 479 | 590 | 734 | 904 | 109 | 132 | 161 | 198 | 242 | 296 | 358 | 431 | 514 | 607 | 709 |

**Fig. 1.7** United Nations population trends, projected to the year 2050, for urban and rural populations globally and in low-income nations

of rural residents, particularly younger generations who seek access to greater opportunity and more modern services than they may be able to access in rural communities (Eshetu & Beshir, 2017). With their departure to cities, the number of rural residents of child-bearing age will decrease, leading to an aging and less robust rural workforce. Though strong urban pull factors among young rural migrants in one study were identified as greater employment opportunities (66%) and a higher level of income (62%), 57% of urban migrants cited the availability of good infrastructure facilities as a prime pull factor for shifting to urban life, placing it ahead of improved social life (36%) and lower risk from natural hazards (46%) as a critical decision factor (Radhakrishnan & Arunachalam, 2017). Regardless of the reasons cited to explain this phenomenon of a global shift toward urbanization, the result is that rural societies across the globe find themselves increasingly separated from markets, resources, economic growth, and even self-respect (Rignall & Atia, 2017). The power that once accompanied an intimate knowledge of rural space and its resources has been completely reversed, and "modern" knowledge and resources that flow through the global marketplace to urban centers have thoroughly disempowered the place-based societies rooted in the countryside (Creed & Ching, 1997).

The loss of ambitious rural youth to urban communities particularly hampers rural societies, most of which are agriculturally based and dependent upon physical labor, and endangers the preservation of place-based knowledge and identity. If we consider the possibility that access to reliable infrastructure could dissuade a portion of these migrants from leaving their native societies, an imperative is placed upon more effectively serving rural communities with appropriate engineering design to retain a country's diversity of indigenous knowledge, land management, agricultural practices, and cultural identity.

The notion of optimization of resources is not a new one and applies as much to the delivery of infrastructure aid to non-industrialized societies as to any other economic endeavor. Because the metrics of development in recent decades have focused on quotas—Millennium Development Goals of 2015, for example, set a 50% reduction in the proportion of people without sustainable access to safe drinking water (WHO, 2013)—the greatest impact on the least use of resources has driven development agencies toward urban interventions. One evaluation of where non-governmental organizations work in Kenya, for example, found that agencies show a strong urban bias when choosing locations to work, drawn by the ease of access, comfort, convenience, and population density as well as by user need (Brass, 2012; Dipendra, 2019).

As a result, the small-scale projects associated with rural communities often fall within the purview of mission trips, service-education organizations, and volunteer groups whose financial resources are limited and expertise is variable (Matthew et al., 2016). Often, these do-good organizations rely on guidance manuals that are not context-specific but offer generic technical guidance and on volunteer Western engineers who draw from their professional experiences in conforming to industrialized-world standards such as U.S. Safe Drinking Water Act standards or EN Eurocodes. Equally frequently, the organizations are associated with, or partner with, agencies that promote particular foci such as religious dogma, capitalistic entrepreneurship, or even market development for particular products. The result? Clashes of rural identity with designer perspective that leads to infrastructure design rejection. Here are two examples of projects I've witnessed during travels and research:

- An indigenous Honduran agricultural community seeking a safe water supply strongly objected to NGO insistence that a nearby spring source be enclosed in concrete, in accordance with government standards for drinking-water protection. The reason? This spiritually focused society has believed for generations that the spring is a supernatural portal into the afterlife, and any encasement of the source would necessitate immediate removal if residents hoped to gain access to eternal existence upon death.
- A Senegalese farming community located far from a natural water source allowed a Peace Corps volunteer to coordinate construction of a hand-dug well and rope pump to ease the burden of village women who collected water for household use. Particular consideration was given to gender equity with the intent of freeing rural women to engage in more economically productive activity. This Muslim community consisted of compounds within whose walls the multiple wives of a single husband were sequestered except when leaving the confines to collect water or work in the fields. Within weeks, the women of the community disabled the pump, expressing a need to regain their freedom to associate with non-family friends while walking the distance to the natural water source.

The infrastructure designers' desires to address overt needs without identifying local context led to a failure of function for the target recipients in both cases, though either infrastructure could have produced steady-flow, high quality water supply. For the Honduras community, globalized standards conflicted with indigenous beliefs. For the Senegalese community, the design failed to consider user motivations and values.

## 1.5 Conclusions

From our examination of the literature in this chapter, we can begin to think about why we undertake engineering projects for populations different from our own, how we assess the performance of those projects, and why rural populations pose both the greatest challenges and arguably the greatest needs for engineering support in the world.

We recognize that stakeholders in engineering projects strive to do good, but the reasons that motivate those stakeholders aren't always as clear or as easily recognized as we may assume. Too frequently the engineers' objectives don't align with those of the users, leading to differing expectations of outcome and objective.

Nowhere is the local context more significant than in the rural communities of the world, which are less exposed to global knowledge, less valued for their own capabilities, and less likely to tolerate technical designs that don't align with their own needs and wants.

But Western engineers aren't educated to think about the users of their designs, nor are we trained to question whether the designs we create for our own communities might be inappropriate and ineffective for people from other places and contexts.

Where does all this lead? First, it requires us to rethink our own role in engineering, our own beliefs about others, and our own implicit biases that lead us to make assumptions about the context in which our technical designs will be used. Like the visitor who cared for the dogs in Calcha, Bolivia, are we providing comfort to others to meet their needs or our own? When we leave, have we created a solution that equips others to thrive in the future, or have we created a dependency that elevates our own power and authority?

These are the questions we'll explore as we move forward.

**Questions for Parable Consideration**

1. What do you think happened to Reina and Brava after the visitor left? Did the visitor give the dogs the nutrition they needed to survive so that they may better forage for themselves, or did she do the dogs a disservice by providing comforts that won't be available later?
2. At what point exactly does the visitor cross the line from helpful to hurtful? Why do we not all agree on this line?
3. What if one of those dogs was your dog, who was temporarily lost? Would that change how you feel about the situation?
4. Is there any sustainable action the visitor could have taken to ensure the dogs' success after her departure? What are the possible unintended consequences of ensuring this sustainability?
5. Why don't the residents of the community feed and care for the dogs themselves? What insight does this provide about their community?

# References

Alverson, H. (1977). Peace corps volunteers in rural Botswana. *Human Organization, 36*(3), 274–281. https://doi.org/10.17730/humo.36.3.k142x53kn6733184.

Amin, J., Denis, J., Harris, B., Ibenegbu, N., Javorszky, M., Maillot, E., & Tripp, S. (2015). *Determining success in community managed rural water supply using household surveys.* Cranfield University.

Bouabid, A., & Louis, G. (2015). Capacity factor analysis for evaluating water and sanitation infrastructure choices for developing communities. *Journal of Environmental Management, 161*, 335–343. https://doi.org/10.1016/j.jenvman.2015.07.012.

Brass, J. (2012). Why do NGOs go where they go? Evidence from Kenya. *World Development, 40*(2), 387–401. https://doi.org/10.1016/j.worlddev.2011.07.017.

Carter, R., Tyrrel, S., & Howsam, P. (1999). The impact and sustainability of community water supply and sanitation programmes in developing countries. *Water and Environment Journal, 13*(4), 292–296. https://doi.org/10.1111/j.1747-6593.1999.tb01050.x.

Chambers, A. (2009). Africa's not-so-magic roundabout. *The Guardian* 25.

Creed, G., & Ching, B. (1997). Recognizing rusticity: Identity and the power of place. In B. Ching & G. Creed (Eds.), *Knowing your place: rural identity and cultural hierarchy* (pp. 1–38).

Dipendra, K. C. (2019). Between rhetoric and action: Do NGOs go where they are needed? *VOLUNTAS: International Journal of Voluntary and Nonprofit Organizations, 30*(6), 1197–1211.

Dunn, E. (2017). Standards and person-making in East Central Europe. In A. Ong & S. Collier (Eds.), *Global assemblages: technology, politics and ethics as anthropological problems* (pp. 173–193). Malden MA: Blackwell Publishing.

Eshetu, F., & Beshir, M. (2017). Dynamics and determinants of rural-urban migration in Southern Ethiopia. *Journal of Development and Agricultural Economics, 9*(12), 328–340.

Ferguson, J. (2012). Global disconnect: abjection and the aftermath of modernism. In J. Xavier & R. Rosaldo (Eds.), *The Anthropology of Globalization* (pp. 136–153).

Grigg, N. (2017). Institutional analysis of drinking water supply failure: lessons from flint, Michigan. *Journal of Professional Issues in Engineering Education and Practice, 143*(3). https://doi.org/10.1061/(asce)ei.1943-5541.0000312.

Haraway, D. (1991). Situated knowledges: The science question in feminism and the privilege of partial perspective. In *Simians, cyborgs, and women: The reinvention of nature* (pp. 183–202). Routledge.

Harvey, P., & Reed, R. (2006). Community-managed water supplies in Africa: Sustainable or dispensable? *Community Development Journal, 42*(3), 365–378. https://doi.org/10.1093/cdj/bsl001.

Hjorth, P., & Bagheri, A. (2006). Navigating towards sustainable development: A system dynamics approach. *Futures, 38*(1), 74–92. https://doi.org/10.1016/j.futures.2005.04.005.

Ika, L., & Hodgson, D. (2014). Learning from international development projects: Blending critical project studies and critical development studies. *International Journal of Project Management, 32*(7), 1182–1196. https://doi.org/10.1016/j.ijproman.2014.01.004.

Imai, K., Gaiha, R., & Garbero, A. (2017). Poverty reduction during the rural-urban transformation: Rural development is still more important than urbanization. *Journal of Policy Modeling, 39*(6), 963–982. https://doi.org/10.1080/15228916.2014.956635.

Krause, M. (2014). *The good project: The field of humanitarian relief NGOs and the fragmentation of reason*. The University of Chicago Press.

Lacy, W. B. (2009). Empowering communities through public work, science, and local food systems: Revisiting democracy and globalization. *Rural Sociology, 65*(1), 3–26. https://doi.org/10.1111/j.1549-0831.2000.tb00340.x.

Lentfer, J. (2017). 'International development' is a loaded term. It's time for a rethink". *The Guardian*.

Lipton, M. (1977). *Why poor people stay poor: Urban bias in world development*. Harvard University Press.

Massey, D., & Massey, D. B. (2005). *For space*. Sage.

Matthew, L., Piedra, L., Wu, C., Diaz, A., Wang, H., Straub, A., & Nguyen, T. (2016). Social work and engineering: Lessons from a water filtration project in Guatemala. *International Social Work, 60*(6), 1578–1590.

Mawhinney, D. B., Young, R. B., Vanderford, B. J., Borch, T., & Snyder, S. A. (2011). Artificial sweetener sucralose in U.S. drinking water systems. *Environmental Science and Technology, 45*(20), 8716–8722.

Mincyte, D., (ed.). (2018). Let's do it! World: Ecological citizenship and the politics of development in Lithuania. Obstacles to development (University of Illinois at Urbana-Champaign). In Sociology of Development Conference.

Nieusma, D., & Riley, D. (2010). Designs on development: Engineering, globalization, and social justice. *Engineering Studies, 2*(1), 29 59.

Radhakrishnan, P., & Arunachalam, R. (2017). Study on factors responsible for shifting of rural youth from agriculture to other occupation. *Madras Agriculture Journal, 104*(1–3), 94–97.

Rignall, K., & Atia, M. (2017) The global rural: Relational geographies of poverty and uneven development. *Geography Compass, 11*(7). https://doi.org/10.1111/gec3.12322.

Shannon, M., Bohn, P., Elimelech, M., Georgiadis, J., Marinas, B., & Mayes, A. (2008). Science and technology for water purification in the coming decades. *Nature, 452*(7185), 301–310. https://doi.org/10.1038/nature06599.

Starkl, M., Brunner, N., & Stenstrom, T. (2013). Why do water and sanitation systems for the poor still fail? Policy analysis in economically advanced developing countries. *Environmental Science and Technology, 47*(12), 6102–6110. https://doi.org/10.1021/es3048416.

Stottlemyer, F. (2017). Personal conversation. (In conjunction with September 13, 2017, International Rural Water Association Board of Directors Video Meeting).

Timmons, A. C. (2021). *Rural-urban contextual data triangulation for international engineering project work*. Master's thesis, University of Illinois Urbana-Champaign.

Vaccari, M., Vitali, F., & Tudor, T. (2017). Multi-criteria assessment of the appropriateness of a cooking technology: A case study of the Logone Valley. *Energy Policy, 109*(2017), 66–75.

Venugopal, R. (2018). Ineptitude, ignorance, or intent: The social construction of failure in development. *World Development, 106*(2018), 238–247.

Wambi, B. (1988). *Domination by cooperation: a third world perspective on technology transfer and information* (IDRC reports, v. 17, no. 1).

What Is Globalization? (2016). www.globalization101.org.

WHO. (2013). *MDG 7: ensure environmental sustainability*. World Health Organization. http://www.who.int/topics/millennium_development_goals/mdg7/en/.

Witmer, A. (2017). Engineering is not development! In Live performance in TED Salon UIUC. Champaign, IL.

Witmer, A. (2018). Addressing the influence of context and development in rural international engineering design. Doctoral dissertation, University of Illinois at Urbana-Champaign.

Woods, M. (2007). Engaging the global countryside: Globalization, hybridity and the re- constitution of rural place. *Progress in Human Geography, 31*(4), 485–507. https://doi.org/10.1177/0309132507079503.

World Urbanization Prospects. (2018). *United Nations Department of Economic and Social Affairs*. http://WUP2018-F03-Urban_Population.xls AND WUP2018-F04-Rural_Population.xls.

# Perspectives, Motivations, and Objectives

**The Parable of the Worms**

An adolescent girl who loves to swim is fascinated by nature. She particularly identifies with aquatic animals and observes that after every rainstorm, a glut of earthworms cover the pavement, often lying motionless at the edge of small puddles in which more worms writhe exuberantly. The girl notices that the worms in the puddles appear to be filled with energy as they wriggle through the water, and without investigating the biological constitution of the animals, she concludes that the lethargy of those worms on dry land results from their lack of access to the water in which the invigorated worms squirm. Quickly, she scoops up the motionless worms and returns them to the puddles so that they may rejuvenate in the environment they so clearly enjoy (Fig. 2.1).

It is only later that she learns the worms at the edge of the puddle likely were resting after escaping drowning, while the worms in the puddles were frantically struggling to reach dry land. Her effort to "rescue" the worms based on her reasoned but incorrect observations actually had doomed them to dying from the very condition they had struggled to escape.

A.-P. Witmer. *Contextual Engineering*. Synthesis Lectures on Engineering, Science, and Technology, https://doi.org/10.1007/978-3-031-07692-3_2

**Fig. 2.1**  An earthworm rests on a leaf after escaping drowning on a rainy day

## 2.1    Why Engineers Do What We Do

From a decade of interviews, observations, and surveys of engineers and engineering students, it's apparent that technical designers believe they participate in international engineering projects because of a sense of responsibility to others (Witmer, 2017, 2018). Project participants typically express a desire to share their knowledge, provide relief, and learn about new cultures, but they rarely exhibit sufficient self-reflection to look more deeply at whether these motivations are genuine. Even when explicitly asked as part of a comprehensive online survey to reflect on their purpose for participating in international engineering projects, many technical designers gave responses like "I want to make the world a better place," "It's the right thing to do," and "I want to help people." While such noble statements are frequently voiced by travelers working on projects, observation of participants at project sites verifies that many project participants also gain personal gratification, whether it consists of seeing new places, accepting the gratitude of the infrastructure recipients, or demonstrating their expertise to others.

This duality of humanitarian motivation and personal gratification appears to manifest more overtly among engineering students, who frequently express their desire to travel while developing engineering knowledge. Stated one engineering student participating in an academic-led project in Bolivia, "There's a focus on giving back using our engineering skills, but it's implicit. We wanted hands-on experience." The student's colleague was even more direct in explaining participation in the effort: "My goal was to gain more practical experience in actual engineering. I didn't see lots of other projects as real-life engineering, and I'm particularly interested in concrete design and concrete pouring." The concrete-channel construction to which the student refers was part of a hands-on experience that the student wouldn't experience in a classroom, he said. As an added bonus, the construction methods used in this Andean region were significantly different from techniques he would learn in the industrialized world, giving him the added advantage of identifying the benefits and drawbacks of alternative methods.

Both students and professional groups frequently exhibit a desire not only to assist societies firsthand through travel and interaction but to demonstrate to others a worldliness and international expertise as a mark of seniority. I witnessed this phenomenon in Guatemala when I encountered a team of travelers from a Midwestern U.S. church on a mission trip, seeking to work with our project team on constructing a water system for a rural community of indigenous Mayans. Many mission group members described themselves as seasoned travelers who had been to the country "three or four times," and their self-representation as humanitarian veterans on a mission to assist the "poor unfortunates" indicated that they were motivated not only by altruism but by the ability to display to inexperienced colleagues their sophistication, local knowledge, and what they referred to as their dedication to "inferior" societies. That demonstration of self-value recurs often among international development engineers. During the same project trip, a young White engineer regularly sought opportunities to demonstrate his familiarity with Latin American culture, insisting he was fluent in the local language (he was not) and navigating local commerce to demonstrate his proficiency in buying local treats. At one point during travel, the engineer insisted on stopping our bus so he could run into a *tienda* (store) and purchase some *Bob Esponja* snacks for himself. This desire to demonstrate his expertise extended to technical design as well. When his inaccurate statements about the impossibility of bacteriological contamination occurring in a groundwater supply were challenged, he stammered and replied that while he wasn't an environmental engineer he had taken a course in fluid dynamics while in college and therefore possessed sufficient knowledge to predict biocontaminant behavior.

When engineers adopt a mien of proficiency, intimacy, and at the same time superiority without becoming an integral part of the local community, they risk deceiving themselves into believing that they know the societal conditions of the user population, even though their understanding of those conditions is filtered through their own experiences of privilege. In fact, the combination of familiarity, affluence, and license-to-serve can create a dynamic of "doing good" at all costs, regardless of the outcome. On one Honduras trip, after taking measurements of the spring source identified to serve a particular community, I informed the team that the volume of water supply a community could receive from this source would be very limited. The funding-agency representative, however, refused to consider whether the project would be feasible, given the lack of water resources, and would not explore its value with the community. My journal entry on that discussion summarized the conversation: "(Engineer 1) and (Engineer 2) understood my concern about undertaking a 5–6 km pipeline down a gorge and back up to the village for 1/2 gallon per minute of water supply, but (the funding representative) got very angry with me and said any water is better than nothing."

In contrast to these engineers and engineering students who perform infrastructure projects in rural areas as volunteers, representatives of NGOs and government agencies who work professionally in non-industrialized societies often make their livings by working with both communities and the volunteer groups that provide funding, expertise, and labor. As a result, these NGO representatives frequently feel obliged to demonstrate the success of past projects so that they may attract funding for future work. As Easterly (2002) states, and my own observations confirm, organizations ultimately engage in "obfuscation, spin control, and amnesia" to demonstrate their value. The display of the rural Honduran water-filtration plant on the funding organization's home page, described in Chap. 1, is an example of how failed performance and engineering errors were irrelevant to the NGO when it wanted to demonstrate its capabilities.

In our Contextual Engineering courses, we refer to this sort of project pattern as "no-consequence design." The technical designer achieves her own objectives of experiencing travel, interacting with another society, engaging partners in funding an impressive-looking infrastructure, learning alternative construction techniques, and demonstrating her own humanitarianism—all with no consequence to her professional reputation or the likelihood of confrontation or litigation from the user population if an infrastructure that fails to perform as promised. In other words, when the engineer is driven by intrinsic motivations, coupled with an ambiguous idea of "helping others," there is no cost to that participant if the operability, durability, or conformity with societal conditions is deficient.

## 2.2  Satisficing the Solution

Let's take a step back and think about why our objectives and motivations are so important to consider. We begin with the term *satisficing,* which is a decision-making process that adopts the first adequate solution rather than investigating further to determine the solution that will optimally address the need. Satisficing is not necessarily a demonstration of technical laziness so much as a common practice in Western engineering to minimize design costs while maximizing the lessons of experience.

An interdisciplinary group of students in a recent senior design class demonstrated well the concept of *satisficing* in engineering. Charged with developing a mechanism to move a large half-hoop greenhouse on a rail from one location to another for a student farm, the controls designers rapidly converged on a solution that quickly proved unmanageable for the shoestring farm operation to address. Rather than returning to the drawing board and investigating additional options, the student team attempted to resign from the project, explaining that the user did not value their recommended solution.

Like these students, many engineers assume that conditions anywhere in the world, working with any people, are invariable and will conform to the designer's own understanding of what works in infrastructure. It also presumes that the design approaches used in the applications with which the engineers are familiar are equally applicable anywhere. In essence, it pre-defines the need based on the designer's experience and converges quickly to a familiar solution. This assumption, driven by the commodification of engineering services in globalized nations, does a disservice to the user population, however, since the needs and conditions of one community almost certainly diverge from the needs and conditions of another. *Satisficing* on design allows designers to ignore those differences so that they may rapidly and effortlessly produce an outcome. If that outcome doesn't satisfy the user population, it is considered a shortcoming of the user and not the engineer.

Nowhere is this error of thought more easily demonstrated than in engineered infrastructure designed by engineers who place great confidence in their own technical knowledge, as well as an enduring belief in the universality of application, regardless of context, location, sophistication, or contradictory advice on relevance. Reliance on industrialized technology is borne out during investigation, design, implementation, and operational phases of infrastructure interventions, and often is manifested in association with the motivational desire to elevate the status of "developing" societies to industrialized standards. This notion of the need to deliver industrialized technology to non-industrialized societies is rooted in the misguided belief that modern technology will better the user society's physical, economic, and societal health. Thus, user context can become for many visiting technical designers a constant of sorts, and the development of solutions can become equivalent in thought to that of the funding source who considered any effort to obtain water worthwhile, regardless of its effectiveness or sustainability of approach.

Consider the use of modern global positioning system (GPS) technology in non-industrialized locations, where the scarcity of satellite signals limits the accuracy of performance. Confidence in GPS, cultivated by its availability in the industrial world on phones, watches, and laptop computers, reassures the engineer that locations with cellphone signals surely will also have precise GPS measurements. GPS accuracy consistently improves with technology expansion, but data precision is still unreliable in many parts of the Earth. On a 2013 trip to Honduras, I watched a professional land surveyor attempt to use a GPS base station and receiver to conduct a topographical survey of a multi-kilometer pipeline path (Fig. 2.2). Though I warned him that the heavy tree canopy and limited satellite accessibility would restrict readings and constrain accuracy, he forged ahead, providing ongoing reassurances that the base station he had brought was top of the line and could function in any environment. The surveyor spent significant time setting up equipment, taking readings, and searching for signals so that he could accurately record distances, elevations, and directions of the proposed path for a water system pipeline. After several days of effort, however, and to the considerable amusement of residents of

**Fig. 2.2** A local technician observes with amusement as U.S. Land Surveyor attempts to use GPS equipment to gather data for a rural community in Honduras

the community, he was able to produce only three readings along the pipeline, none of which were inflection points, inlets, or outlets.

Reliance on existing data, such as a student engineer's favorite tool—Google Earth— also provides an unwarranted level of confidence among industrialized-world engineers working in non-industrialized locations. Engineering students often rely on topographic data collected online without exploring the literature, which clearly states that tool resolution is insufficient for engineering use. In the case of Google Earth, measurement resolution is 30 m in the vertical direction and 90 m in the horizontal direction (El-Ashmawy, 2016), which is horrifically insufficient to design a gravity-fed water system. Yet I've seen students turn again and again to the web to answer questions that they otherwise wouldn't know how to answer, confident in the reliability of the data they gather.

If you believe, as many do, that there are some ground truths about people, nature, and the world that can guide us in design, permit me to introduce you to the community of El Guarango, Ecuador. This lovely village lies near the Ecuadoran coastline and is home to some of the lushest fruit farms in the world. Dragon fruit, papayas, cacao, melons, passion fruit, coffee, and a range of exotics not known to U.S. palates … the list of produce grown there is staggering, though the community has no access to reliable water supply and farmers truck in tankers and bottles of water for use in both home and farm. If you look around the town, you notice that many of the homes appear ramshackle, built of stripped bamboo and roofed with corrugated sheet metal (Fig. 2.3). Once in a while, you come across a home built of masonry (Fig. 2.4), which stands out amid the wooden structures. If you ask a group of Western engineers to list out the first words that come to mind when looking at the bamboo houses, invariably you'll get a list that includes "rundown," "poverty," "unstable," and "disheveled." Show them the masonry home, however, and the majority of engineers will list words such as "sturdy," "middle/upper class," "safer," "expensive." It's always a shock to these engineers when they learn that both houses boast electricity, a refrigerator and stove, a flat screen TV, and comfortable furniture. In fact, in El Guarango, the bamboo house is very much the preferable form of construction because it can flex and sway through the frequent earthquakes that strike this area, allow cool breezes to flow through the interior during sweltering days, and be easily and cheaply repaired should anything damage it. The masonry houses, meanwhile, often are replacements for the previous homes that collapsed during a major earthquake,

**Fig. 2.3** A home constructed of stripped bamboo in the community of El Guarango, Ecuador

**Fig. 2.4**  A masonry home in El Guarango, Honduras

and many of their residents have spent considerable time sleeping under tarps because they couldn't afford shelter until the government reconstructed their dwellings. Besides the rigidity of the materials, these houses are stiflingly hot, difficult to keep clean, and far less easily repaired should they be damaged.

Western engineers often make erroneous assumptions about the people living in these two types of houses because of their own lack of familiarity with local conditions. Untrained in exploring the local context, many industrialized-nation engineers will create a structure that satisfies their own understanding, relying on familiar materials, ignorant of weather, and neglectful of community tradition and identity.

One might assume from this analysis that Contextual Engineering is anti-technology. Nothing could be farther from the truth. But the application of technology without consideration of user context can produce outcomes where the technology itself, much like the objectives and perspectives of the technology designer, misaligns with user desires and capabilities.

The Las Mesas water-filtration plant introduced in Chap. 1, which employed modern filtration processes but failed to produce safe water within days of start-up, provides a superlative example of how an introduced technology not only can fail to address a community's needs but instead can create hardship and resentment from that community toward the donor organization, the local partner NGO, and infrastructure designers themselves. Though the NGO provided cursory training to system operators, the dedicated *fontaneros* did their best to maintain the quality of the water. When we visited the plant a year after it was placed into service, a test of water quality indicated that finished water leaving the plant was equivalent in turbidity and bacterial contamination to the raw influent entering the plant. Examination of the infrastructure indicated that modern equipment and materials were used in the plant construction, but filter maintenance—which requires pumping capability and an electricity source—was not even considered in the operation, making the plant not only inoperative but hazardously bacterial in a very brief time.

Some engineers have cited the expression of interest in modern technology by user populations as a justification for providing unfamiliar and often complex equipment, even if it creates a dependency or results in a limited operational lifespan. It feels good to elevate the user population's technological access, some international engineers say, by stretching their capabilities and better aligning them with the industrialized world. This introduction to industrialized-society standards is not lost upon some user populations, who occasionally express admiration for Western devices and consider access to them to be a sign of societal advancement. My journal records the words of one rural Cameroonian village elder as he sought to extract additional support from a student group that had just completed construction of a community water system:

> The leader of the village said he expects our help so that they can obtain electricity that will allow them to watch television day and night. Many referred to their village as Young America. I was stunned, as was the rest of the team.

In reality, sprinklings of Western technology are not uncommon in non-industrialized societies, often proudly displayed as a symbol of global ascendency that has been delivered by outside organizations or local politicians who seek to leverage support in exchange for amenities. The contrast between flat-screen televisions or smartphones and the simple devices used for cooking, cleaning, farming, and sanitation can be stark in some non-Western communities.

In rural India, where many residents live in homes of thatch with dirt floors and openings rather than doors or windows, electrical service lines penetrate the roofs to power television sets that were provided by the government to disseminate propaganda. A conversation with one woman from the rural area demonstrates how industrialized technology has woven its way into the fabric of life in the community, even as residents were reliant upon a contaminated borehole to obtain water for drinking, cooking, and cleaning:

> The sanitation group interviewed a young woman with a 3-year-old daughter and an 8-month-old son. She described how she fell in love and became pregnant, but because her husband's family objected to her, she was 8 months pregnant when they married. She now lives with her husband, children, and in-laws. When she described her day it was a shock! She arises and cleans, wakes and bathes the children, then watches soap operas on TV. In the afternoon, she naps, prepares dinner then – amazing – watches more TV.

Some international aid practitioners debate whether such technology enhances the lives of non-industrialized societies or contaminates them with Western values. At a mission house in the Lake Atitlan region of Guatemala, I met a number of U.S. expatriate volunteers who often gathered for coffee and healthy debate about the appropriateness of technology introduction. One particularly passionate nun there objected strenuously to the U.S. aid-driven construction of a bridge for a rural indigenous community because it would alter their bucolic way of life by introducing vehicle traffic, product delivery, and consumerism. Others in the conversation challenged her by saying that it also would allow residents to access transit, thereby expanding their opportunities to work outside the community, as well as introducing the ability to ship in foodstuffs that would expand their diet and improve nutrition. The debate continued without resolution throughout my two-week stay with the mission, illustrating the dichotomy of impacts: preservation of indigenous lifestyles versus enhancement of living conditions to improve health and welfare.

Such debates are common among aid workers in indigenous societies, with those opposed to introducing globalized technology citing the loss of indigenous identity and knowledge. The debates sometimes go to such extremes that in one case, an Ecuadoran NGO insisted on promoting locally appropriate technology for an indigenous community even as the community itself had become cosmopolitanized and identified more strongly with metropolitan Quito than with its historical indigeneity. The NGO worker recruited Western organizations to assist in developing an irrigation system for the community's farmlands, which had experienced severe drought for several years due to changing weather patterns. When another engineer and I met with the ruling *cabildo* (elected community leadership) regarding these initiatives, we were bluntly told that there was no interest in farming and residents would prefer we help them to develop an eco park on the land to promote tourism. Even with this directive from the community itself, the NGO continued to push for farmland irrigation, reportedly lobbying community members to challenge the *cabildo* to withdraw its alternate request for assistance so that her own image of a static and quaint ancient society could be preserved.

While many aid workers routinely question their own role and impact in non-industrialized societies, though, engineers are prone to plunge forward without a thought about why they're working with a particular community, what that community may actually want or need, and how they may represent those wants ambiguously because they value the support of industrialized partners. Why do we comfortably insert our technology into other locations without fully investigating the conditions and needs of the user as well as the impact of our own actions?

Tradition in engineering is a particular challenge that extends not only to technology confidence but to the very identity upon which the engineering profession was built. A study by professors at Ohio State and Kansas State universities (Burack & Franks, 2004) that looked at the suppression of diversity in engineering found that Western white males strove to protect the elitism of professional engineering by excluding those who are not male, Caucasian, or from the industrialized nations. To justify their superiority, then, engineers dismissed people and the technologies they espoused as lesser if they didn't conform to the norm. This would suggest that the engineering profession itself, not consciously but through decades upon decades of self-definition and elitism, discourages us from self-reflecting on the questions that are so very important to Contextual Engineering practice. Perhaps this is why engineers often find it so very difficult to ask a user population—often not male, not Caucasian, and not industrialized—for their opinion about what solution is needed, how it should be delivered, and whether it is appropriate for the people who will use it.

## 2.3 The Perspective of the Infrastructure User

We've talked a lot in this chapter about the perspectives of engineers who work with a population outside their own experience, but what about the perspectives of the user population? Travel to a remote non-industrialized community and ask them what they need and how they'd like to get it, and you'd best be prepared for the startled looks you'll get in response. I've asked that question again and again while traveling through Central and South America, Africa, and Asia, and the response from communities has been universal: "No one has ever asked me that before!".

To explore the user perspective, let's first overlay Wambi's words from Chap. 1— "(Technology) is encoded with the characteristics of the society which developed it, and it tries to reproduce that society"—with the recognition that our own societal characteristics promote wealth, materialism, and power. If we introduce Western technology elsewhere, then, are we promoting our own values to societies that don't share them, all with the promise that we'll improve their lives? Is that not the very definition of Colonialism?

If that's a daunting thought, here's one that's even more discouraging. When our goal is to learn from a user population what it wants and needs, we may discover that simply asking the question isn't always going to generate a clear and useable response. Much the way each population is unique in its values, beliefs, capabilities, and needs, each population also is unique in how it approaches interactions with others. And here's where Contextual Engineering can get really complicated.

Consider an example of how much engineers can learn about a user population by talking with and learning from them while working on an infrastructure design with them. In this example, the very nature of thought that resides within a population can shape the way it approaches an engineered design. And while that nature of thought may not directly affect the function of infrastructure, it certainly could change the way the community interacts with the infrastructure to meet their own needs.

In the indigenous region of Guatemala where Kaqchikel, a Mayan dialect, is spoken as frequently as Spanish, people feel a strong spiritual connection to nature and view life as something that unfolds before them in union with nature. During a visit to the community of Nueva Providencia, Guatemala—at the time a recently established community of refugees from natural disasters and exploitive relationships with the coffee industry—I

**Fig. 2.5** A Kaqchikel woman in rural Guatemala who valued her ability to own a home but lacked the language to describe how water accessibility could affect her life

spoke with a woman (Fig. 2.5) who for the first time in her life was able to own her own home. Because she spoke only Kaqchikel, her words were translated by a man from the village into Spanish, which was then conveyed to me in English by my colleague, "R".

> R and I visited one woman in her home, a poorer home made of wood. The woman was nursing a baby in a sling as she talked with us. The house was startlingly small – two beds and a little more space in a single room for a family with seven children. The next building was a lean-to for cooking, *la cocina*. A fire was lit to make tortillas. R asked the woman her feelings about how having water at her house will change her life. She had trouble answering, and looked very puzzled as the translator and R tried several ways of asking the question. Finally, R explained that the Mayan language doesn't allow for comparatives or conditionals like "may be" or "better than" because the simplicity of life is either "do" or "do not," "have" or "have not."

This woman could not conceive of how her life would be altered because she lacked the words and the thought process to consider the impact of an alteration upon her life. She could not react, could not verbalize a response, to questions about what she would want from a water system because she simply didn't have the language to form an understanding of the question and reply. Until our conversation, though, this was irrelevant because no one has asked her opinion about the system, whose construction was nearly complete. Assumption of a society's attitudes about water or sanitation or agriculture or transportation, then, is a risky business unless the infrastructure designer purposefully familiarizes herself with that society. A Westerner might assume, for example, that most economically poor societies simply haven't the resources to care for household animals like dogs or cats, and conversations with individuals from societies as distinct as Cameroon and Guatemala demonstrate that there is a great deal of puzzlement over the industrialized-world predisposition to pamper pets. And yet, depending upon the community, regardless of economic standard, attitudes toward animal companions can be strong even if not overt. During my work with an extremely poor rural farming community in the Dominican Republic, for example, it was common to see mongrels trotting among the houses, sipping water from puddles, and approaching visitors to see if they had any food to share. Most residents would shoo the dogs without warmth—throwing rocks at them, swatting them with sticks, charging at them aggressively to send them away. During one trip on the back of a pick-up truck to assess conditions at a spring site 5 km uphill from the town, though, one of the residents observed a dog chained to a tree and asked the driver to stop. The dog had been chained there for several days, the man said, and clearly was dehydrated and dying. Though he barely had the resources to feed his own family, he insisted that his friends help him to remove the dog from the tree, taking great care to protect themselves should it attack, and place it in the back of the truck. The resident cared for the dog and attempted to restore it to health, though colleagues reported it did not live long. By cursory observation, one could assume that dogs were considered nuisances in the community, unworthy of human attention and resources. Yet the determination of the resident and his friends to rescue a suffering animal indicated that some people valued compassion for living creatures, regardless of their value to the household.

If values and motivations are not always overtly displayed, sometimes the technical designer must dig deep to understand contextual conditions that can affect the nuances of infrastructure design for an unfamiliar society. As mentioned briefly in Chap. 1, I learned of *los ojos* during an assessment trip to an indigenous Lenca community in Honduras. At the time, I was accompanied by several students who welcomed an offer by a village leader to see the *ojos*, a term applied to a particular type of spring found in the mountains near the village. As we walked through woodland, the village leader pointed out several trees, one of which was particularly poisonous, and told the story of a man who had constructed a house of one of these trees and then fell ill, only recovering after he begrudgingly went to the forest and apologized to the tree stump for his transgression. The leader then pointed to some undergrowth and said to watch for a particular plant that would mark the presence of the *ojo* (Fig. 2.6). When we reached the spring, I did some quick water sampling and admired the vigorous flow, suggesting that this could be a resource of high volume and high quality to supply the village system in the future. But when I mentioned enclosing the spring with a concrete catchment tank, which is a standard for public water supply not only in the U.S. but in Honduras as well, the village

**Fig. 2.6** An *ojo*, or sacred spring in El Tablon, Honduras, is found by looking for specific flora that grows nearby

leader quickly terminated the visit and said the spring could not be covered. "This is a spring that is guarded by a snake, and you cannot put concrete near such a spring," he explained tersely as he led us back out of the forest. It was only after returning to the U.S. and conducting research into Lenca beliefs that we learned why the village leader reacted so strongly. By Lenca tradition, springs are a portal between the earth and the afterlife, guarded by a snake who guides souls through the passageway. Had the group followed Honduran design standards and covered the *ojo* with a concrete catchment, the community would have demolished it, fearful of otherwise facing an eternity trapped on Earth.

These examples demonstrate that it's no easy task to understand the perspectives, motivations, and objectives of a user population. Conditions may lie below the surface, maybe unvoiced, or may simply be taken for granted within the community even though outsiders don't share the understanding of its significance. But the examples illuminate how objectives and drivers can be very different between the engineer and the community she's working with, providing a peek into one explanation for why technical designs can miss the mark in addressing the needs of the user population.

How do we cope with this? Appropriating the famous Donald Rumsfeld quote of the 2000s, an engineer easily can identify the known knowns of a user community—physical conditions, natural resources, governmental oversight, for example. The known unknowns can either be determined or disregarded—things like resource demands, infrastructure footprints, population size. But the unknown unknowns—those areas that can only be determined through a clear understanding of the community, require patience and comfort with the inquiry. They also require the engineer to be open to exploring the place-encoded technologies, regardless of whether they were developed by university-trained technical experts or community practitioners with no formal education, because these technologies can provide insight into the people who created them.

One final note on the unknown unknowns that should be considered—In Contextual Engineering courses, we refer to an example of the Embarrassing Aunt as a phenomenon that can keep the unknown unknowns hidden forever. Here's what we mean: say you want to bring your new partner home to meet the family for the first time, and you want them to make a good first impression. But your opinionated aunt, who lives next door, has a dreadful habit of popping in when someone visits so she can share embarrassing stories about your childhood and argue politics with your parents. She's your aunt and you don't want to disavow her existence. But you don't want her to knock on the door when your partner is meeting family for the first time either. And so you convince her to go to a movie the night you're coming by, just so she doesn't scare off your new partner.

Every population of people has its weaknesses, its embarrassments, its flaws. Often, communities don't want to lie to you about their existence, but they don't want to parade them before you either. And so some of the critical unknown unknowns, those perspectives or motivations that lie in the dark but can affect the outcome of infrastructure implementation, can be sent off to the movies while you're visiting the community with which you intend to work. This isn't deception so much as a convenient avoidance of sharing uncomfortable information. And it exists everywhere. These are the unknown unknowns that have the potential to create conflict or to lead to disregard for the infrastructure you design with your user population. And while you may never meet the Embarrassing Aunt during your interactions with the user community, you need to recognize that she's very likely there.

## 2.4    Conclusions

Every stakeholder in an international engineering project brings along their own set of perspectives, motivations, and objectives. Some stakeholders may be transparent in their motivations; others may not even be aware of the root motivations that drive them to participate in a project.

Engineers have been taught for generations to be confident in our own capabilities. We rely on our own experience to tell us the quickest, most effective solution to an infrastructure need. But when working with populations that are outside our own experience, this reliance on experience can lead us to *satisfice* a solution rather than deeply exploring the options.

Engineers must be very self-reflective to identify all the reasons they participate in a project as well as to acknowledge the biases and predispositions that guide their design decisions and interactions with other stakeholders. The profession may have imposed some of those reasons upon them. Other reasons may be self-generated, based upon the engineer's own life and professional experience.

The voice of the user population rarely is heard, in part because Western engineers tend to devalue non-industrialized knowledge and in part because the complexity of place-based thought and language may get in the way of the engineer's understanding.

**Questions for Parable Consideration**

1. Why was the adolescent girl predisposed to believing that worms like swimming in puddles? Were her intentions pure?
2. What actions or thoughts might the girl have employed to cope with the guilt of killing works she sought to save?
3. Were her actions unreasonable, or were they founded in logical conclusions formed from observation?
4. How does this parable related to performing technical design for populations that we may observe but not fully understand?
5. What might you do differently when assessing a user population in terms of infrastructure need, based on this parable?

# References

Burack, C., & Franks, S. (2004). Telling stories about engineering: Group dynamics and resistance to diversity. *NWSA Journal, 16*, 79–95.

Easterly, W. (2002). The cartel of good intentions: The problem of bureaucracy in foreign aid. *Policy Reform, 5*(4), 223–250. https://doi.org/10.1080/1384128032000096823.

El-Ashmawy, K. (2016). Investigation of the accuracy of google earth elevation data. *Artificial Satellites, 51*(3), 89–97. https://doi.org/10.1515/arsa-2016-0008.

Witmer, A. (2017). Personal Travel Journals. Bolivia, Cameroon, Dominican Republic, Ecuador, Guatemala, Honduras, India, Nigeria, Senegal: 2006–2017. Handwritten diaries, unpublished.

Witmer, A. (2018). Addressing the influence of context and development in rural international engineering design. Doctoral dissertation, University of Illinois at Urbana-Champaign.

# Standards, Self-sufficiency, and the Rural

3

**Parable of the Plunger**

A group of 20 engineering students and a faculty member are visiting their client community in rural Honduras, inhabiting a village house together while exploring physical and societal conditions that will inform their water system design. Late one evening, a graduate student quietly tells the faculty advisor that the group's sole latrine, a bucket-flush commode, has become plugged and will not flush. Together, the two engineers investigate the situation and begin to search for a toilet plunger, which cannot be found on the property. Envisioning a hazardous late-night drive to the nearest town to buy a plunger, the faculty advisor tries one last thing and asks a fluent Spanish speaker from the group to run to the nearest home and ask to borrow a plunger. The student soon returns, followed by the neighboring home's patriarch, who wearily carries his ubiquitous machete as he tucks his pajama shirt into worn khakis.

When the patriarch determines the toilet is indeed plugged, he walks behind the guest house and begins hacking at a sapling with his machete, leaving the engineers puzzled. When he returns, he holds a sturdy shaft of freshly hewn wood in his hand, and he earnestly studies the stray litter that has become entangled in a barbed wire fence surrounding the property (Fig. 3.1).

He says nothing as he extracts a ragged plastic grain sack, quickly wraps the sack around the sapling shaft, and gestures to a student using a roll of duct tape for his own craft to tear off a length. Taping the contrivance together, the neighbor enters the latrine and plunges the toilet clear with two swift thrusts. The technical "experts" among the group witness this ingenuity with shame, embarrassed that their proposed solution had been to drive many kilometers to purchase a ready-made plunger that would serve the same function as the sapling innovation.

© The Author(s), under exclusive license to Springer Nature Switzerland AG 2022
A.-P. Witmer. *Contextual Engineering*. Synthesis Lectures on Engineering, Science, and Technology. https://doi.org/10.1007/978-3-031-07692-3_3

**Fig. 3.1** A sapling-shaft
plunger innovatively devised to
unplug a Honduran toilet

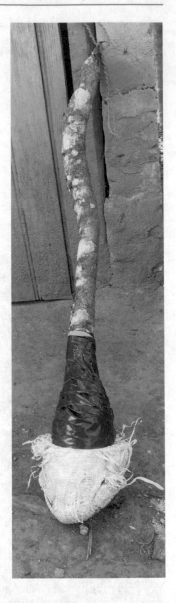

## 3.1   The Standard of Applying Standards to Engineering

It is a foundational hypothesis of Contextual Engineering that a stronger understanding of the interrelationship between local (often indigenous) technical practices and globally influenced design expertise can allow for the selection and implementation of more effective and sustainable technological interventions that match user needs on a community-by-community basis. This understanding becomes strained, however, when

Western engineers are educated to believe that advanced technology, created through rigorous research and safeguarded by industry standards, can learn nothing from place-based practices and indigenous knowledges. We discussed at length in Chap. 1 the sociological impact of standards when applied to societies whose conditions and context were not even considered in their establishment, but standards nonetheless remain at the forefront of Western engineering training and Western practice, regardless of the context of the population to which the standards are applied.

In this chapter, we explore more closely why Western engineers promote their own standards and dismiss the contextual standards of non-industrialized societies, how these engineers often dismiss the technical knowledge and capabilities of non-industrialized societies, and where this disparity is most readily observed.

Let's start with thinking about the conditions that lead us to place so much value on industrialized-world standards. The social theory of Technological Determinism asserts that all humans follow the inevitable advancement of technology, which will progress in a particular direction regardless of societal conditions or intervention. This theory has pervaded engineering education for more than a century (Choi, 2022), and appears to be widely embedded in the thinking of both engineering educators and students. While the determinist theory has been challenged in recent decades, and alternative theories that incorporate societal agency into technology progression have arisen (Feng & Feenberg, 2008), determinism still seems to have a tight hold on Western engineering-education thought, implicitly driving students toward focusing only on the science and expecting society to deal with whatever outcome results. A notable consequence of this enmeshing of determinism and engineering education is that many Western engineers are empowered to consider through determinism that technology advancement is universally applicable. In other words, all societies, because they're part of the human family, must accept Western technological advancement as an inevitability because determinism guides its evolutionary path.

Let's take a second to discuss the meaning of the "evolutionary path." If you subscribe to Technological Determinism—which we don't!—you believe that technical processes will march on and society will trot along beside them to keep up or lag behind and ultimately become lost. More simply put, you're either technologically sophisticated and can embrace the advancement of Western technology, or you're doomed to extinction. And yet we know that many non-industrialized societies have embraced alternate value systems, alternate economic bases, and alternate objectives to those of Western populations. This would suggest that there is a different evolutionary path that may neither value nor conform to the influences of ongoing technical development.

Take, for example, some of the Quechuan communities of the South American Andes, which give little consideration to global markets or currency, instead valuing self-sufficiency and communal support. These communities possess knowledge of global commerce, but they choose not to covet such things as high fashion, lush lawns, or professionally groomed pets. These populations can boast a rich ancestral history of technical achievements—their ability to sustainably farm lands that most Western agriculturalists would consider barren is nothing short of stunning. So why would Western engineers impose cutting-edge Western technologies upon these populations and dismiss as irrelevant the alternative technical knowledge path—the technical *evolutionary* path—that the Andes Quechuan communities have followed? Because our determinist indoctrination leads us to believe we are obligated to advance any society's development by bringing to them the very newest innovations from the top minds in the world.

The sociological literature (e.g., Dunn, 2017; Ika & Hodgson, 2014; Haraway, 1991) provides a persuasive argument that engineered infrastructure design should not be fused with societal "development" efforts if it is intended to address a non-industrialized community's physical needs. Infrastructure is not and should not become leverage for social change. Its purpose is only to address a physical need.

Now that we understand the influence of Technological Determinism, we can see why there's a concatenation of engineering and development into a single effort when industrialized-world engineers work with non-industrialized societies. One of the strongest tools in promoting this conjoinment is reliance on international standards, which unintentionally (I hope!) debases local values, skills, and predispositions. We can see why this coupling of engineering, development, and standards-making for non-industrialized populations seems so natural to so many Western engineers. When we think of the deterministic motivation to promote Western technology in non-industrialized societies, we can't help but recognize that it serves a second purpose: we're not simply providing a technical solution to address a user need, but we're *developing* a population in our own image. The next time you catch yourself talking about designing a technical solution and needing to "educate" user populations to discard place-based practices so that they may conform to globalized standards for infrastructure design, construction, and operation, know that you're employing the Technological Determinism with which your education has been imbued. And intentionally or unintentionally, you're uniting the activities of engineering design with dictating a new evolutionary path for the user population.

## 3.2    The Value and Risk of Engineering Standards

A fundamental requirement of engineering education, as determined by the influential Accreditation Board for Engineering and Technology (ABET) is a comprehensive understanding of, and the ability to apply, discipline-specific industry and engineering standards and codes. While ABET accredits more than 4,300 engineering education programs in 41 countries, those accreditations are noticeably absent in broad swaths of non-industrialized societies, including the majority of the African continent and significant portions of South and Central America. ABET is not the only vehicle of engineering education accreditation in Western nations, though. In parts of Europe and Australia, for example, agreements with resident accreditation programs allow graduates to be considered as being in accordance with ABET education. This is not so for the majority of non-industrialized countries, which still maintain accreditation programs for professional engineering education, but which vary in criteria and often place no emphasis on design or industry standards. Nonetheless, by sheer numbers alone, ABET maintains a global standard for requiring engineering students to learn the "appropriate engineering standards."

Based on one institution's recent accreditation review, ABET reviewers will explicitly demand evidence that industry standards are cited and incorporated into course materials so that students are fully apprised of the application and advancement of best-available practices for engineering. What does this mean in the engineering classroom? A senior design course taught in one department of this accredited, elite engineering program included projects serving U.S. Fortune 100 industries, regional specialized industries, and non-profit service providers in both the U.S. and non-industrialized countries. In accordance with ABET, students in the course were required to identify the relevant standards and apply them to design for each user population, regardless of their type of project or user population location. For the Fortune 100 company, which provides products worldwide but focuses its market on industrialized nations, this requirement was simple. But for the project teams performing engineering design with a non-profit working in a U.S. indigenous territory or for an academic institutional client in West Africa, the application of "appropriate standards" became much more challenging. Are U.S. Safe Drinking Water Act standard operating pressures critical when working with a native American population whose water system had not been maintained for several decades and who lack the financial resources to invest in system improvements? Were Occupational Safety and Health Standards for ferry rails necessary when designing a low-cost boat for the West African school that would allow agricultural students to move farm equipment from school buildings to fields across a river? (Before the ferry, students occasionally were swept away by raging river currents while trying to wade across the river with equipment in hand.) Examples such as this demonstrate that standards are indeed as context-determined as values, physical conditions, and preferences. As the examples from the literature above demonstrate, though, Western engineers are trained to identify "appropriate standards" as those that apply to their own living situation, and even when the situation changes, the appropriateness of the Western standards is retained.

In fairness, engineering standards are critically important when placed in the appropriate societal context. As a public water supply engineering designer in the Midwestern U.S., I relied upon design standards, from state administrative codes to Ten States Standards to American Water Works Association Standards to Safe Drinking Water Act standards, not only to ensure the quality of design but also to guide me through numerous interrelated technical considerations so that nothing was left to chance. Without these forms of guidance, I could not have done my job effectively, and if I didn't adhere to these standards I could have been held professionally and even criminally liable for any negative outcome. So I do value my knowledge of Western engineering standards and I do draw upon them when working with non-industrialized populations. But do I insist on adhering to them for every project I encounter around the world? No.

How, then, may a Contextual Engineer parse the conundrum of when to apply globalized Western standards to conditions they encounter elsewhere, especially when they simply appear not to be relevant? As my students in Contextual Engineering will tell you, the answer I give to many questions such as this is quite unsatisfying: "it depends." It depends on the capabilities of the user population, on the resources at hand, and on the availability of alternative approaches that will address the need while doing no harm. Scholarship is virtually silent regarding the notion of assessing the applicability of standards, which places the onus upon technical designers to educate themselves and suspend their own technical predispositions to better align with local standards and practices that will produce the greatest user acceptance without creating undue risk.

## 3.3   Innovative Self-sufficiency

Inherent in any discussion about engineering design in non-industrialized societies and the application of standards to technical design is the question of what knowledge already resides in a place that could help to answer design questions. Because many non-industrialized communities lack access to formal education, the indigenous knowledge of these communities is perceived as valueless, substandard, or unsophisticated by Western standards. Take, for example, the conversation I once had with an eminent environmental engineering educator, who dismissed the concept of contextual knowledge with these words: "I'm not going to build something out of mud and twigs because it's 'appropriate technology'," the professor said, using air quotes. "That's a cop-out. I want something that will work and will last." I recall those words each time I look at the photos of the Ecuadoran houses from Chap. 2, which demonstrate that mud and twigs sometimes *are* the superior materials, functioning longer and more strongly than concrete masonry.

As engineers educated in the deterministic-technology tradition, we too often assume our own technology and technical knowledge are superior, which leads us to dismiss the innovations and skills that reside in the non-industrialized world. And yet, if it's true that technology is encoded with the characteristics of the society which developed it—clearly, Bruno Wambi was *not* a Technological Determinist!—why are we loathe to recognize that the technology which develops with people in a place is best suited to address the unique needs of that population?

This concept first caught my attention in graduate school while taking an interdisciplinary course in Ecohydraulics, which spent a great deal of time considering the unique functions of fish varieties based on the water body in which they reside. A most simple example would be the rainbow trout versus the lake flounder. One would not expect a flounder to fare well in a fast-moving current because of its small tail, its flat round body, and its predisposition to hiding in sand or mud to await passing prey. The sleek, powerful bullet-shaped trout, though, thrives in fast-flowing rivers where it may dart about or quickly rise to the surface to catch its prey. Could you tell a trout to behave like a flounder or vice versa? Of course not. The capability of each fish evolved out of the place in which it lives, and it functions best by leveraging the conditions of that place to swim and eat.

Technical capability is no different. Residents of the African Sahel possess agricultural skills that are simply untranslatable to the mountains of the Andes. As absurd as it would be to ask our Midwestern corn farmers to plant a steep slope by hand, it's equally absurd to suggest that tractors could improve agriculture on an indigenous terraced farm in Guatemala, where farmers tie off at the top of their land and rappel down the slope to care for their crops. Similarly, an infrastructure inspector seeking to examine the underside of a bridge deck in New York City could not employ the skill of the Honduran farmer who, upon realizing he needs to climb to the top of a 2-m water tank, hacks down saplings with a machete and constructs a wooden ladder in minutes. Access to materials, as well as possession of the proper tools, limit technical capabilities, and those conditions often are dependent upon the environment, the societal practices, and the cultural experience of the individual.

This is place-based knowledge, and it grows out of and continues to develop in non-globalized societies throughout the world. Technological determinists may dismiss this knowledge as mud and twigs, as primitive practices that are unworthy of preservation. But if you spend enough time in non-industrialized communities, learning from and talking with their residents, you come to appreciate the unique value of place-based knowledge and the significance it could hold not only in preservation and application to the local community but in the edification of Western designers who may gain insight into processes and applications they'd never before considered.

A review of my travel journals and experiences in many different rural societies showed that this valuation of place-based knowledge and appreciation of its complexity is not a reflection of a technical process or a design so much as it is a way of thinking or, as Wambi said, an encoding of the characteristics of the society that developed it. In Contextual Engineering, we describe this place-based technical knowledge as *Innovative Self-Sufficiency*, or the ability to craft a solution by considering conditions and available materials. It is a critical component of survival and advancement in the non-industrialized world, and insistence on replacing it with manufactured and imported equipment and materials threatens the existence of the thought process itself.

While acknowledging the resourcefulness, economy of effort, and use of place-based materials and considerations as a type of knowledge that should be preserved, one should not imbue Innovative Self-Sufficiency with the belief that it is a meticulous application of technical expertise. At times, Innovative Self-Sufficiency can take the form of piecemeal repairs and haphazard modifications, leaving it subject to criticism that solutions are slapdash and don't produce long-term functionality. Pipeline repairs in a Dominican Republic community, for example, demonstrate that non-sustainable solutions often are employed to prevent a 5-km pipeline from clogging with vegetation (Fig. 3.2) or to patch damage

**Fig. 3.2** Netting is thrown over a water supply intake pipe, which also has been filled with rocks to prevent detritus from entering and clogging the pipeline

to PVC pipe from falling rocks on a hillside (Fig. 3.3). These solutions were employed by untrained local residents who observed the flow of water to the village declining, first because of leaves and vegetative detritus entering the pipeline and creating a clog, then because of leaking and broken pipes. While they were effective in addressing the community's immediate needs, they lacked the technical rigor to ensure the successful operation of the pipeline for the foreseeable future. This same community's lack of technical expertise led at least one resident to earnestly propose an irrigation solution that consisted of hand-pouring volumes of concrete between the walls of a narrow rock canyon at the top of the mountain on which a spring flowed, thereby trapping millions of cubic meters of water and creating a reservoir that could be tapped indefinitely. It was a creative idea, but one that any structural engineer could tell you would be doomed to almost immediate failure because of hydrostatic pressures, potential for undercutting, and a host of other vulnerabilities. Inarguably, the expertise and technical knowledge that is cultivated through a rigorous engineering education can be brought to bear on an infrastructure solution, but it should not ignore nor disregard the preservation and incorporation of local innovation. The attuned technical designer must be alert to the presence of Innovative Self-Sufficiency and incorporate it into a rigorous design so that it matches the understanding and expectations of the user population. In an Eastern Guatemalan community,

**Fig. 3.3** Bicycle innertubes are used to patch failing PVC pipe in the Dominican Republic

for example, where changes in weather patterns and deforestation of mountainsides for cookstove fuel have resulted in significant maize crop failures, agronomic engineers affiliated with a local university are working with the indigenous population to expand its repertoire of crops while continuing to support its heritage as corn farmers. For one farm in this region, the farmer has elected to plant among his maize fields crops of both sugar cane and cypress trees, which flourish in the highland climate and provide cover for the coveted coffee trees that can produce significant cash crops. Processing of the sugar cane is difficult and labor-intensive, yet the farmer—with the help of supporting local engineers—quickly innovated a sugar cane press that he constructed using logs and a trough fashioned from a sheet of galvanized metal (Fig. 3.4). The device extracts sugar juices for processing and functions almost identically to the massive mechanical extractors used in the industrial sugar factories I visited in India. Simple and constructed completely from local materials using the farmer's own mechanical intuition, this press allows him to produce sugar products at a fraction of the cost that would be associated with purchasing a mechanical press. This melding of Innovative Self-Sufficiency and technical support from the agronomic engineer produced a solution that did not conflict with the farmer's own self-sufficiency, yet it expanded his capabilities on his own terms, creating a win–win solution that introduced an appropriate technology without sacrificing traditional values associated with working with the gifts provided by the land.

**Fig. 3.4** A handcrafted sugar cane press mimics sophisticated sugar-processing equipment using rudimentary tools and materials

## 3.4    Where Standards Threaten Innovative Self-sufficiency Most—The Rural

We talked in Chap. 1 about Why The Rural Matters in Contextual Engineering. Non-local volunteer organizations that serve rural populations from the West often are smaller, less experienced in international service, and more likely to implement a design that conforms with their own experience and standards. When those designs clash with local values, identities, and beliefs, and fail to acknowledge and incorporate the Innovative Self-Sufficiency that resides there, the resulting infrastructure often doesn't function for long. This frequently leaves residents disheartened and distrustful of outside support, which can then disqualify them from future projects that have the potential to be more sustainable. So with the greatest of intentions, engineers working with rural populations typically fail to recognize that their adherence to industrialized-world technology and standards actually limits their ability to perform effective and sustainable infrastructure design. The particular temptation to trust in the existence of a deterministic application of technical processes results in devaluing a user population's local power and agency to address its individual needs effectively and sustainably using Innovative Self-Sufficiency. The impact of the engineer following this thought process has ramifications that reach far beyond technology into the very fiber of the non-industrialized society.

You'll recall from Chap. 1 that globalization has promoted increased urban migration and abandonment of rural lifestyles and communities, which poses a particular threat to the preservation of place-based, indigenous identity, knowledge, and practices. Particularly the youngest, most ambitious, and most able residents of rural areas have been identified as moving to cities to gain greater access to reliable infrastructure, hold better jobs, and live in greater comfort. As they are compelled to leave their rural lives behind for the promise of opportunity and ease, the identities and knowledges that diversify the global society diminish and are lost to future generations, rarely respected or acknowledged by those who work with these communities because of the devaluation of non-Western technical capabilities.

Meanwhile, NGOs and government agencies seek to achieve the greatest impact for the largest number of beneficiaries at the lowest possible cost (Krause, 2014), which by definition leads to infrastructure interventions in the most populous cities in a given society. Additionally, because of globalization's impact, cities are more readily connected to world-normalized standards of behavior, consumption, and values, meaning that they are more likely to adopt an infrastructure design created by a globalized technical designer relying upon Western standards that disregard local knowledge. These are the reasons that urban migrants often cite when leaving rural society. The globalized identity that lures people to cities also makes the implementation of a place-indiscriminate infrastructure more feasible. But in rural societies, where the local still dominates over the global in terms of both practices and values, ready acceptance of world-standard technologies is less prevalent. And economies of scale for infrastructure interventions do not exist in agricultural communities, where residents have spread far apart and centralized systems and equipment can be prohibitively expensive to install. The result is a self-perpetuating cycle of decreasing rural population, which leaves rural areas even less desirable as beneficiaries of support from humanitarian agencies and organizations (Satterthwaite et al., 2010), even as declining agricultural productivity that accompanies the flight of farmers decreases food security for societies already struggling with availability of affordable local nutrition.

Contextual Engineering actually offers a unique understanding of rural communities to improve engineered infrastructure for those often perceived as the most difficult user populations to satisfy—those who are culturally entrenched in a non-globalized environment, those whose experiences clearly differ from the experiences of a Western technical designer, and those whose agricultural activities support not only their own livelihoods but the food security of their home country as well.

Before we move into techniques for practicing Contextual Engineering in the next chapters, we take a moment to acknowledge that we have spoken about community context with an implication that for non-industrialized populations, there is absolute homogeneity among community members with respect to identity, belief, value, and capability. Nothing could be farther from the truth for, of course, anytime a collection of humans comes together, there is diversity of thought, action, and belief. As we'll see in the coming chapters, we often are challenged to identify a dominant or prevalent characteristic to assess a population, but this does not mean we have to assume everyone conforms to that characteristic. When does this matter? When it is irrelevant? Listen for the Contextual Engineering students' groans at the unsatisfying answer: "It depends."

## 3.5    Conclusions

Technological Determinism is a commonly taught belief that technical progression is independent of societal wants or needs. Though the theory of determinism is widely discredited (Choi, 2022), engineering education continues to focus on the advancement of technology independent of societal influences or impact.

Engineering standards provide guidance to technical designers on how to communicate minimum performance expectations, set parameters for risk and reliability associated with a technology, and assist engineers in identifying resources for best-practice design processes.

The application of Western standards to populations that follow a different evolutionary path may rely upon an implicit endorsement of Technological Determinism. As a result, Western designs for non-industrialized populations may intentionally or unintentionally conjoin engineering with development for the purpose of bringing along societies that may neither desire to follow nor believe in the evolutionary path being imposed upon them.

Populations on an alternate evolutionary path from Western society often possess their own place-based knowledge and technology, known as Innovative Self-Sufficiency. Innovative Self-Sufficiency may possess superior applicability to Western standards or may reflect conditions that indicate Western technology will not be maintained. The designer must acquaint himself with local Innovative Self-Sufficiency before judging its appropriateness to address a need.

Innovative Self-Sufficiency is sustained and promulgated most frequently in rural areas that are untouched by globalized influences. Place-based knowledge and ancestral practices, in the form of Innovative Self-Sufficiency, may contribute to a stronger understanding of technical needs and capabilities and should be identified and explored, rather than discarded and abandoned in favor of Western innovation.

**Questions for Parable Consideration**

1. Putting yourself in the shoes of the patriarch and using only items in the room/around your house, how would you construct a solution?
2. Can you describe a community where the innovative sapling plunger would be an inappropriate solution? Why would this be the case?
3. Why do you think people may be willing to innovate solutions for some things but not others?
4. Does performance of an innovation solely determine its technical value? Why or why not?
5. For what reasons would you be most likely to dismiss the sapling plunger as inferior engineering? Is it related to materials? Is it related to complexity of design? What other reasons might contribute to dismissal of the plunger as a technology?

# References

Choi, H. (2022). *Understanding the development of professional social responsibility among engineering students from the perspectives of the critical theory of technology* [Unpublished doctoral dissertation proposal]. University of Illinois Urbana-Champaign.

Dunn, E. (2017) Standards and person-making in East Central Europe. In: A Ong, & S Collier (Eds.), *Global assemblages: technology, politics and ethics as anthropological problems* (pp. 173–193). Malden MA: Blackwell Publishing.

Feng, P., & Feenberg, A. (2008). Thinking about design: Critical theory of technology and the design process. Philosophy and Design, pp. 105–118. Springer, Dordrecht.

Haraway, D. (1991). Situated knowledges: The science question in feminism and the privilege of partial perspective. Simians, Cyborgs, and Women: The Reinvention of Nature. Routledge, pp. 183–202.

Ika, L., & Hodgson, D. (2014). Learning from international development projects: Blending critical project studies and critical development studies. *International Journal of Project Management, 32*(7), 1182–1196. https://doi.org/10.1016/j.ijproman.2014.01.004.

Krause, M. (2014). *The good project: The field of humanitarian relief NGOs and the fragmentation of reason.* Chicago, IL: The University of Chicago Press.

Satterthwaite, D., Mcgranahan, G., & Tacoli, C. (2010). Urbanization and its implications for food and farming. *Philosophical Transactions: Biological Sciences, 365*(1554), 2809–2820. https://doi.org/10.1098/rstb.2010.0136.

# Context in All Its Glory

**4**

**The Parable of the Sanitation Program**

During a trip to Guatemala to work with a humanitarian organization supporting a poor mountainous community near Chiqimula, a group of engineers and fund administrators sat in an outdoor café, talking with a representative of a Northern European development organization that desired to promote sanitary latrine construction in rural areas. The representative described in detail the program he was proposing to the funder—experts provide construction training to select community residents, who then share their knowledge with neighbors. Such a training program would cost the foundation a substantial fee for the trainer's time and materials, the rep said, but the investiture of knowledge in the community would lead to rapid adoption of sanitary waste infrastructure. The funder asked what he would have to show at the end of the year to justify the investment, and the rep testily responded that the community would have knowledge. "That's not what I mean," the funder responded (Fig. 4.1).

"How many latrines would have been built?" as a result of investing in the program. The rep, growing angry, responded, "That's not how development works!" When the observing engineer began to laugh at the conversation, the two men turned to her and asked why she was amused. Her response: "Unfortunately, that is EXACTLY how development works. But I find it funny that the two of you are arguing about how to assess a project's success when the people whom you seek to help aren't even a part of the conversation." Days later, the fund administrator returned to this comment, still shaking his head and saying, "I can't believe we were arguing about the value of this project and the community wasn't even there to share their thoughts!"

© The Author(s), under exclusive license to Springer Nature Switzerland AG 2022    57
A.-P. Witmer. *Contextual Engineering*. Synthesis Lectures on Engineering. Science.
and Technology. https://doi.org/10.1007/978-3-031-07692-3_4

**Fig. 4.1** The village of El Durazno sprawls over a hillside, at the bottom of which the community's water source lies susceptible to up-gradient contamination

## 4.1    Why Context Matters

As we've explored in the preceding chapters, respect for place-based knowledge and skill should combine with technical expertise to generate an infrastructure intervention solution that is both appropriate and rigorous, and the engineer who integrates this knowledge is far more likely to produce an outcome that addresses user populations' physical infrastructure needs sustainably. The investigation of place-based characteristics is a key attribute of Contextual Engineering. Because the foundations and basis of Contextual Engineering are unmapped in either the sociological or the engineering literature, we create a definition for this term:

> **Contextual Engineering,** *noun*
> The creative application of science, mathematical methods, societal understanding, and place-based knowledge to address a physical need that serves the user of the innovation while recognizing the influence of stakeholder motivations, capabilities, and values.

There is no mention of development or humanitarian objectives in this definition. Contextual Engineering exists for the sole purpose of improving the technical design to address an identified need for a user population, regardless of that population's economic state, market isolation, or cultural identity. In other words, Contextual Engineering legitimizes the application of the social sciences to technical design for the purpose of improving design functionality and sustainability. We've shown numerous examples that demonstrate why technical proficiency alone is not sufficient to address a user population's needs, particularly if it is a community that lies outside the pervasive influence of globalized information, advertising, entertainment, and products. Creating a recognition that societal understanding and indigenous knowledge are as critical as mathematics and science to the success of technical design for non-industrialized societies will encourage the designer to employ rather than discard place-based Innovative Self-Sufficiency, focus on local needs rather than global standards, and engage with the user population as a technical designer rather than a redeemer. In the Contextual Engineering approach, user education and development are secondary to design implementation, because context determines the existing capabilities as well as the propensity for evolving those capabilities and internalizing information exchanges before an infrastructure is deployed.

In applied terms, Contextual Engineering challenges the engineer to pose the question "*why?*" in addressing a physical infrastructure approach before addressing the traditional design question of "*how?*" Creating an understanding of why an infrastructure is best suited to the user population's needs offers the extra step for designers of familiarizing themselves with place-based conditions, knowledge, and influences in an effort to identify the best technical approach that will meet user population constraints, both technological and societal.

## 4.2   What Does This Mean for the Designer Perspective?

We've discussed how Western engineers are taught to consider technology as deterministic, and the newest technology is always an inevitable advancement that must be applied to all populations if they are to advance as well. We've also debunked the value of Technological Determinism by placing it into the framework of alternate evolutionary paths, defined by the location, the people, and the time in which a need resides. But how do we convince the designers who are so deeply conditioned to regard technical advancement as separate from the society that, in fact, the user society is the key element to determining the appropriateness of the technology itself?

Let's state it bluntly here:

> Engineering effectiveness is inherently dependent upon designer recognition that the technical approach must conform to the context of the user and not to the designer's own technical comfort level.

This recognition does not come without anxiety, in that it requires designers to cultivate awareness of their own motivations and objectives (Krause, 2014) as well as of their

**Fig. 4.2** Model of the
contextual levels of perception

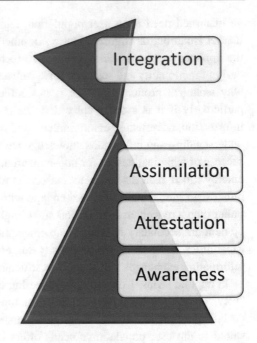

predispositions toward judging others' needs and values through the filter of their own
experiences. The capability of an engineering designer to produce a usable technology or
infrastructure rests upon the ability to incorporate user conditions without bias and with
full awareness of the gaps in understanding that inevitably will remain.

Contextual Engineering uses a model to identify the level of perception with which an
engineer views an unfamiliar user population. Figure 4.2 presents the model, which was
developed out of research into how designers assess their user population and the impact
it has upon their ability to engage context with technical design (Witmer, 2018).

Let's spend a little time with this model to better understand the contextual perceptiveness that an engineer should seek to cultivate in working with a user population. The three
A-team perceptive levels—Awareness, Attestation, and Assimilation—are considered the
productive progression in understanding a need and context of an unfamiliar population
(Witmer, 2015). The fourth level, Integration, emerged during research and occurs when
a contextual investigator immerses in the user population to a degree that results in loss
of self.

Let's place this hierarchy in the framework of Contextual Engineering to better
understand the significance of perceptive level. The first level—Awareness—is the acquisition of basic knowledge that allows an engineer to identify that needs exist and may
be addressed through a physical infrastructure intervention for a user population. The
engineer also is aware that the user experience is different from his or her own.

In the second level—Attestation—the designer witnesses the user population and its needs, either in-person or virtually to a degree that the designer gains some sense of understanding of the conditions of the context. Firsthand exposure to user context, for example, may assist the designer in building an understanding of the need and potential interventions, but that understanding is framed within the designer's own experiences and belief system, rather than illuminating for the designer the underlying core values and conditions that govern the user's true experience.

It is only through the third level—Assimilation—that the designer can begin to build a more complete understanding of user-community needs, adopting a perspective that disregards the designer's own experience. Assimilation allows the designer to identify and incorporate place-based capabilities, beliefs, and values into the technical infrastructure without dominating interference from his or her own experience and value system.

If you're thinking "no big deal…just spend a little extra time and you'll assimilate" you're still stuck in an attestative perspective! Assimilation not only is difficult to achieve, but it requires a great deal of effort to retain, since as humans we're most comfortable falling back on our own knowledge and experiences when encountering the unfamiliar. Especially when that unfamiliar condition conflicts with our own value set or beliefs, we're prone to running back to the safety of our foundational values and beliefs. And yet, resisting that flight toward the familiar and instead adopting an assimilative perspective not only gives the designer a clearer glimpse into the needs of the user population, but also serves the purpose of creating a relationship that places both technician-consultant and user-client on equal footing by deprofessionalizing the consultant (Jones, 1990) to elevate the user's status. It's through this relationship that a collaborative design may be developed. The accomplishment of assimilation requires significant self-exploration on the part of designers to be able to recognize when they begin to slip into attestative mode and impose their own belief systems upon the design.

Before we talk about the fourth level of contextual perception, let's take this theoretical discussion into the real world with an example that illustrates the value and risk of each level in the A-team of perception.

A painting hangs in my living room, which I share in Fig. 4.3. It's not a great painting. In all honesty, it's a pretty mundane pastoral scene. But it holds deep meaning to me, primarily because I have an assimilative understanding of its value. Can you guess why?

Let's start with what you know by looking at this painting. From it, you may assume this barn actually existed somewhere at some point in time, and if you're particularly sharp-eyed, you may be able to determine where it was located by the type of trees overhanging the structure and architectural style of the barn itself. You are aware of the barn, and if I tell you this barn is important to me, you know that as well. But that's the extent of your knowledge. Could you identify from this awareness why this barn is so significant that a painting of it hangs in my living room? Of course not. All you know is that it's a barn.

**Fig. 4.3** A bucolic barn in the woods (painting by Frank J Washabaugh Jr.)

So you want to know more. OK, let's go visit this barn, which at the time the painting was created was located on a remote rural farm tucked deep in the woodlands of northeastern Connecticut. You walk into the barn and note the old wooden construction. You observe the oil stains on the floor, no doubt from farm equipment that had once been stored there. You observe the three steel metal milking stanchions and the small steel corral in the back corner, as well as the overhead doors and sliding barn doors on both front and back walls. You see the ladder to the loft and maybe you climb it to see that the upper floor is indeed a loft. There's still hay dust drifting in the air so you can assume it was functional at one time, but if you walk across that floor, you may notice that the floorboards are weak and can barely handle your weight. If you're really astute, you might walk around the outside of the barn, notice the tangles of wild black raspberries growing behind and beside it, and maybe even spot the cellar door that leads to a bubbling artesian well in the foundation of the building. You can attest to the smell of the barn, the features of the barn, and the way light shines through the large sliding doors on the loft to hit the cracked floorboards. With this attestation, can you identify why this barn painting is so significant to me? Of course not. You've explored the barn and applied your understanding of it, but this doesn't tell you why it matters.

So you take that next, uncomfortable step and ask me why I care so much about this very ordinary New England barn. This verges on the touchy-feely, which you'd rather not do, but it's necessary to understand the value of the painting. I tell you long, colorful stories of an idyllic childhood playing in this barn, swinging on the stanchions, watching my brothers play basketball in the loft, and laughing when a foot would break through the floorboards. I talk about gorging on the black raspberries with my sister, and watching my father pull an ancient cultivator out of the front overhead door to plow his garden. I tell you how my father and I worked together to convert this barn into a horse stable, chopping down trees from the woodland behind it to make the split rail fence that we built to create a paddock in the foreground lawn. I describe sitting on the ledge of the square second-floor door, waiting for the hay tractor to pull up and the workmen to toss the hay and straw bales through the large top doors with exquisite strength and accuracy. Then I explain the painting, which my father made while recuperating from his first heart attack. I mention that he died a few years later of his second attack, while I was still a child, and my family moved away because the farm was too difficult to maintain without him. Hmmm, maybe you are beginning to understand the value of this painting now?

I can assure you that I have not shared everything that makes this barn so very special to me. Some things are for only my family and me to know. Some things are embarrassing in that they show my own human frailties. Some things are less relevant than others, or so I believe, so I simply don't share them. But with your questions and respectful curiosity, I disclose the key bits of information that place the barn painting into context and provide insight into what I value, what I treasure, what I fear, and what I'm capable of doing and being as a consumer of your technical design.

This example should demonstrate that the task of developing contextual perception is akin to acknowledging that your own experience and perceptions do not give you a look into someone else's reality. It's a journey, not a checklist, that leads you to sufficiently understand your user population's needs and capabilities. First, you become aware of your intended users. Then you witness their conditions, recognizing that you are applying your own understandings and experiences to the process. Finally, you calibrate your understanding to the user population so that you can explore needs and capabilities without judgment. That is the assimilative perspective of engineering that Contextual Engineers strive to achieve.

We had tabled our discussion of the fourth level of perception that was shown in Fig. 4.1 until now. When evaluating data compiled as part of the formulation of the Contextual Engineering concept, a curious condition reoccurred whenever investigators spent a great deal of time embedded within their user community. The diagnostic tool we use to identify contextual conditions frequently spits out skewed results for the long-term investigators when compared with the consistency of their colleagues who looked at the same community, leading to the hypothesis that an additional level of perception may exist—integration. As shown in the figure, integration is an unstable perspective in that it allows investigators to step outside their own experiences so much that they view the user population from within the set of beliefs and conditions that reside within the population—usually reflecting the perspective of those people with whom the investigator most closely associates. Remember that we acknowledge the heterogeneity of belief in any collection of people. So if the investigator gravitates to one segment of the collection, their understanding of the totality of the community may create a new limitation on the designer's ability to address the needs of various sectors of the user population. In essence, it substitutes the investigator's new identity for their previous preconceptions, potentially applying blinders to the assimilative view through overexposure.

## 4.3   Variability of Objectives and Motivations Among Stakeholders

This condition acknowledges the differences that exist within a user population. But how about the differences that exist among the stakeholders working on an engineering project intended to address the needs of a population. Are there differences there as well? You betcha!

Let's assume you're embracing the Contextual Engineering concept now and have calibrated your own perspective to the assimilative level. You're not the only decision-maker in the design process, though, so how do you calibrate to all the other decision-makers? Does the funding resource spend equal time in considering the impact of their perspective and motivation for participating? Does the NGO? For that matter, does the user community itself? As we recognize all the baggage we carry into a project with regard to our predispositions, biases, and root motivations for participation, we must also recognize that *every* other participant carries baggage as well, and more often than not, our collective luggage is not a matched set.

We're going to be realistic here and say that just as it isn't the role of the Contextual Engineer to change the way its user population approaches technology and infrastructure, it's not your role to change the way the other stakeholders think either. But it is your role to recognize they may think differently, act differently, and even participate for different reasons. Once you've acknowledged this fact, you've made progress in uniting the stakeholders in a collaborative effort. But recognition that each participant may be engaged in the design process for their own reasons and with their own expectations only means that the Contextual Engineer must be vigilant at all times to prevent collisions of objectives or mismatches of beliefs by watching, questioning, and negotiating with all parties.

## 4.4   The Influences on Design Appropriateness

This discussion of perception and observation gives us a reason to rethink the way we perceive our user population and co-stakeholders when designing technology, but what relevance does this have to actual technical decision-making? After all, sophisticated scientific knowledge, rigorous mathematical calculations, and deliberate equipment imple-mentation are critical to creating a robust, sustainable technology, and there's no argument that the most contextual engineer in the world will fail if they lack technical rigor. But it's been demonstrated in case study after case study that technical rigor clearly is not the sole determinant of a satisfactory outcome in infrastructure design, since its implementation is prejudiced by non-technology influences of human relationships, desires, competencies, and values. How can one break down those influences into a clear set of constraints to which technical decision-making is applied?

Engineers have tried to parse the sociological influences on technical effectiveness for decades, primarily using case studies to identify wealth, power, and education as non-technical conditions that can alter the way a user population interacts with the technology provided to them (note that I say "to them" here, not the preferable term of "in partnership *with* them"). These three conditions may be more succinctly summarized as economic, political, and educational influences. The literature also indicates through sociological evaluations of the impact of globalization on local societies that a tension forms between local identities and global standards. Thus, Contextual Engineering recognizes that a cul-tural influence may be related to the adoption of a technology, particularly when it is provided by Western engineers. Finally, case studies acknowledge that outside engineer-ing service providers frequently fail to identify user willingness or comfort in operating a technology, maintaining it, or evolving it to meet their own needs and capabilities. This could be characterized as a mechanical influence governed by a user's ability to work with and sustain a process with which it's comfortable. Other conditions and conflicts discussed in the literature, from language differences to religious tolerances, can neatly fall within the five influences described here, and so Contextual Engineering uses these five key influences to determine the appropriateness of a technical design for a specific user population.

Contextual Engineers continue to explore the meaning of these influences, incorporat-ing studies in economics, sociology, engineering, political science, and geography, as well as direct interaction and discussion with user populations in South and Central America, sub-Saharan Africa, South and East Asia, and indigenous territories of North America. The literature discussed herein (e.g., Massey, 2005; Otte, 2013) also has informed the selection of key critical influences, into which other considerations such as environment and equity, may be incorporated. From the insights gleaned in all these investigations, we summarize the Contextual Influences thusly:

- Political influence addresses the influence of power within a society. High political influence suggests that power dynamics are a significant feature of that society, and the failure to recognize those power dynamics within a population and incorporate them into the technical design can create conflict and rejection by sectors of the population.
- Economic influence addresses the influence of need, and can be unrelated to mone-tary wealth. High economic influence indicates that a population does not consider itself to have the resources to meet basic needs and feels great stress when confronted

with additional demands upon its limited resources. This influence can confuse casual observers of the Contextual Engineering process, because one intuitively may assume high economic influence equates to high access to wealth, when in fact the opposite is true.

- Cultural influence addresses the values and identities that are specific to the community as a whole. This by no means suggests that any culturally influenced community will be homogeneous in belief and identity, and in fact, the most culturally heterogeneous community can still be highly influenced by cultural considerations. A high cultural influence within a society, however, indicates a very strong adherence to a shared value, identity, or fundamental belief, which must be recognized and respected in engineering design.

- Educational influence reflects the desire of the population to learn new concepts and processes as well as to share new information with others. It does not reflect the overall level of education within a community so much as the value that residents place upon formal or informal acquisition of knowledge, even if that knowledge may conflict with inherent beliefs and experiences that reside within a context.

- Mechanical influence considers the capability of a community to keep things running. Much the way a skilled mechanic can adeptly and often creatively keep a machine operating without gaining new knowledge or modifying function, a population with high mechanical influence can apply resourcefulness and energy to make things work.

In Contextual Engineering, these five influences combine to provide a technical designer with a greater understanding of the context in which technical design development must be framed—the assimilative view screen. The unique combination of relative influences that each user population produces is called the Contextual Fingerprint, which we use to inform technical design decisions.

In the next chapter, we'll talk more about how to identify the Contextual Fingerprint through directed investigation, and in Chap. 6, we'll explore how the fingerprint can assist technical designers to produce a technology that most closely fits with user population needs, beliefs, and capabilities.

## 4.5    Conclusions

Contextual Engineering is the creative application of science and place-based knowledge to address a physical need that serves the user while recognizing the influence of stakeholder motivations and objectives.

A Contextual Engineer must recognize that any technical product or process must conform to the context of the user, rather than to the intent of the designer, if it is to be maintained and operated effectively.

Identification of user context relies on the calibration of the Contextual Engineer's perception to the assimilative perspective. While perceptive levels of Awareness and Attestation are necessary steps in building engagement between the designer and the user population, it is only when the designer seeks to understand the value of a technology from the user population's perspective that contextual understanding may be achieved through Assimilation.

Five key influences determine the Contextual Fingerprint of the user population, which governs the adoption, operation, and sustainability of a technical infrastructure or process. Those influences consider the population's power dynamics (political), ability to meet needs (economic), desire for new knowledge (educational), shared identity (cultural), and organizational facility (mechanical).

**Questions for Parable Consideration**

1. How could you appease the funder who wants a physical and countable result while maintaining the integrity of a contextually appropriate design?
2. What would you do if you learned that one of the stakeholders in a project is motivated by views that you fundamentally or morally oppose? How would you navigate this?
3. How might the stakeholders in this project engage with community members more effectively to align their own objectives while addressing user population needs?
4. Do you think stakeholders in the latrine project are all motivated by a single objective, regardless of what they indicated when talking? What might be the impact of exploring primary, secondary, and even latent objectives associated with each stakeholder?
5. Consider a project you have worked on for which different team members had different goals. What were those goals? Why did each stakeholder hold them? Did everyone's goals align, or were there conflicts in agreement?

# References

Krause, M. (2014). *The Good Project: The Field of Humanitarian Relief NGOs and the Fragmentation of Reason*. The University of Chicago Press.

Jones, B. (1990). *Neighborhood Planning: A Guide for citizens and Planners*. Routledge Publishing.

Massey, D. (2005). *For space*. Sage.

Otte, P. (2013). Solar cookers in developing countries; What is their key to success? *Energy Policy, 63*(2013), 375–381.

Witmer, A. P. (2015). Letting go for success in international service. Live performance at TEDxUIUC, Champaign, IL, April 19.

Witmer, A. P. (2018). The influence of development objectives and local context upon international service engineering infrastructure design. *International Journal of Technology Management & Sustainable Development, 17*(2), 135–150.

# Investigating Context

**The Parable of the Ultraviolet Treatment Donation**

In the late 1990s, an organization of U.S. water professionals committed to providing engineering support to a rural Guatemalan community, with whom it became acquainted through a relationship between one of its members and a former community resident. Professional engineers went to work creating a catchment system in a nearby river that collected surface water and piped it to homes, but they failed to account for the variability of flows and water quality in the river associated with seasonal rains. Not long after the system was operational, they began to receive reports that during the rainy season, river water was spilling over the catchment filter and entering the system with the consistency of chocolate milk. Heavy rains and eroded soils overwhelmed the catchment, which had been designed during the dry period when engineers were most comfortable visiting the rural region. A professional organization member—a manufacturer's representative for water-treatment equipment—donated two state-of-the-art ultraviolet disinfection units to the project, earning high praise from the organization. These units, which use high-efficiency electronic lamps inserted into pipe-flow chambers to disinfect water as it flowed past, were valued at $40,000 USD each at the time and considered to be a cutting-edge treatment technique that was only beginning to emerge as best available practice for disinfection in Midwestern U.S. public water supplies.

The engineers, however, struggled to apply this generous donation to the project for two reasons: (1) disinfection using ultraviolet light requires a water supply with low turbidity, meaning that it would not effectively treat waters containing high suspended organic and inorganic solids; and (2) ultraviolet lamps require electricity, which was not available in this community at the time.

A.-P. Witmer. *Contextual Engineering.* Synthesis Lectures on Engineering, Science, and Technology, https://doi.org/10.1007/978-3-031-07692-3_5

## 5.1    Investigating Context

It's one thing to acknowledge that context exists and is relevant to engineering decision-making. It's another thing altogether to identify context, especially when we've already demonstrated that context is an elusive knowledge that often doesn't reveal itself, at least until trust and camaraderie with the user population have been established. For all the rigorous training that engineers undergo in the sciences to allow us to solve a problem, rarely are we challenged in our education to draw upon empathy and humility to ask the questions needed to *identify* the problem we're seeking to solve. And here lies the rub.

Perhaps one of the greatest frustrations for engineering students who begin to study Contextual Engineering is that they aren't able to dive right into the information that makes them most comfortable—the physical conditions and mechanical operations that must come together to make an infrastructure, product, or process work. Instead, they're challenged to think about their own context first, then they're asked to reflect upon the conditions that underlie the societal context of the user population—the set of societal, physical, psychological, and theological conditions that have played out over time to create the current context in which the engineering need resides. This is a tremendously daunting task, particularly for students who have been conditioned to believe in Technological Determinism, devaluing the relationship between technical design and societal conditions. But it's not insurmountable.

Contextual Engineering has developed a methodology, which we call the 3-4-5 method, to build an engineer's understanding and sensitization to user context, and then to investigate that context's relevance to technical design. You've heard individually about some of the phases associated with this method, but we'll put them together here into a process you can apply.

First, though, the stipulations:

- Will Contextual Engineering explorations give you insight into every possible condition that resides in a population and may affect the use, maintenance, and adaptability of a technology? Not a chance. Just as you'll never know everything about the barn painting shared in Chap. 3, you'll never know everything about your user population. But you *will* acknowledge the existence of unknown unknowns so that you can remain more agile in your design.
- Is this methodology a checklist of things to do and questions to ask? No sirree! This methodology allows the engineer to build a sensitivity to the activities and explorations that can lead to a stronger understanding of the community, but no technique can provide a foolproof process for solving every known unknown and identifying every unknown unknown.

With these conditions in place, let's take a look at the 3-4-5 method and how you can use it to better understand the needs of the user.

## 5.2 The 3-4-5 Method

The name of this method refers to the Three Levels of Perception (the A-team perceptive levels), the Four Quadrants of Context, and the Five Influences that Govern Technical Appropriateness (the key contextual influences). We'll discuss in detail here how to apply these concepts to engineering investigation for the purpose of building a contextual understanding of the user population's needs and conditions.

### 5.2.1 Advancing Perception

As we discussed in Chap. 4, a Contextual Engineer strives to achieve an assimilative understanding of the user context both through self-reflection and through active interaction with the user population and collaborating stakeholders. But how does one accomplish this? We do this through a workshop that examines the types of bias to which humans are subject, the inherent weaknesses of human observation, and the predispositions that guide us toward what is familiar before considering what is not. Sherlock Holmes says we have to eliminate everything that is impossible and whatever remains, however unlikely, is the reality. So our goal is to learn to think like Sherlock by refining our own perceptiveness to eliminate those conditions that don't exist in a population, allowing us to see the reality that remains, whether it aligns with our own context or not.

Before a Contextual Engineer approaches an investigation for the first time, he or she must undergo a one-hour calibration workshop to understand how easily we allow our cognitive biases to govern our understanding of the world. Activities that we use to explore our own biases include videos that demonstrate how unreliable our own perceptions of reality can be when we expect or look for something specific, discussions of the cognitive biases that we rely upon to simplify our thought and guide our understanding; examinations of how we interpret photos of people based upon our own experiences and perceptions, and activities that demonstrate how prone we are to aligning with the perceptions of others—even if we don't agree—because humans are hardwired to avoid conflict and seek conformity. Let's break this down a little further.

Most of us have seen a video or been subjected to a live-action experiment in which something unexpected happens quickly and we're asked to describe what we saw. When I was a freshman in a communications course many, many years ago, the professor had a man run into the lecture hall, threatening to shoot the professor, then run out. Once we all calmed down, we learned that this was an activity designed to test our perceptions and we were asked what we saw. The majority of the 200 students in the classroom collaborated to describe the man's face, clothing, gun, and voice as he shouted at the professor. The incident had been surreptitiously recorded, and when the recording was played back, the room full of aspiring journalists was ashamed to learn the man actually brandished a banana, which he waved quite overtly so we could all see it. Such an activity would never be undertaken today (and we strongly discourage trying anything like this on a university campus!) but it demonstrated that ten-score students were convinced they saw something nonexistent because they were not prepared to observe, did not make note of observations when they were self-absorbed with abject fear, and convinced themselves and each other of what they thought they had observed in the wake of the event. As one of the terrified students who was sitting very near the front of the lecture hall, I can attest that I saw a gun, not a banana. And yet I was wrong. This demonstrates that reality and the *perception of* reality can be very different animals, and while we can't entirely change what we perceive as reality, we can recognize that it very well may be an illusion promoted by emotion, influence, or predisposition.

One of the purposes of the calibration workshop is to address the topic of predisposition. Just as we recognize that context lies within the user population for whom we perform a technical design, we must recognize that context resides within us as well, and it results from the experiences that have shaped us throughout our lives. This context formulates for us a set of cognitive biases—not malicious so much as functional—upon which we rely to fill in the blanks of our understanding. This is why someone from the northern U.S. looks at a house made of stripped bamboo and declares it inferior to a masonry house—if we lack the understanding of atmospheric conditions in Ecuador, we fill in the blanks and assume a bamboo house would be drafty. Similarly, when I tell colleagues that I am traveling to La Paz, Bolivia, in February, they congratulate me for going to warm climates because they know that Bolivia is in the southern hemisphere and enjoying the summer months. They may lack the understanding that La Paz is the world's highest-altitude national capital and typical temperatures there hover around 35 °F at night and 65° at midday all year long. Filling in the knowledge gap with an awareness that the month of February is summertime in the southern hemisphere and many northern U.S. vacationers go south in winter produces an inaccurate understanding of the Bolivian context.

Cognitive biases have been well studied and cataloged into complex and countless forms that guide our thinking, our understanding, and our gap-filling in thought and action. A lifetime of learning and practice would not eliminate the influence of all our cognitive biases, but building a comfort level with questioning your perceptions and examining whether a bias could be influencing them is the outcome we strive to achieve in calibrating our perspective to achieve an assimilative understanding of our user population.

### 5.2.2 The Four Quadrants of Context

Once our Contextual Engineers at least acknowledge the existence of variability in perception and understanding, we move forward in examining the conditions that determine the context of the population we're investigating. To do this, we draw from the Contextual Engineering Daisy Model, which was developed to understand many (but not

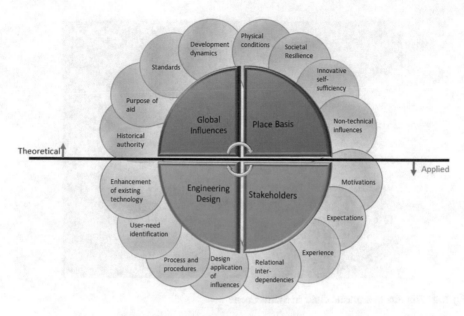

**Fig. 5.1** The Daisy model of contextual engineering, which divides investigations into four quadrants

all) of the considerations that must be explored because of their relevance to technical design decision-making. Figure 5.1 shows the Daisy Model, which identifies the four quadrants of contextual consideration as Global Influences, Place Basis, Stakeholders, and Engineering Design. This may be summarized more succinctly as the Global (outside or pre-existing conditions), Local (physical and societal conditions of place), People (predispositions and relationships associated with stakeholders), and Process (conditions of design, functionality, implementation) quadrants, which together comprise the "4" of the 3-4-5 method.

This investigation is intended to bring the Contextual Engineer forward in her understanding of the user population by examining a variety of conditions through both personal interaction and analysis of secondary data. Those data may include news stories, social media posts, project applications, conversations with residents and neighbors, stakeholder explorations, and other investigatory processes, all with the purpose of building a stronger understanding of the four quadrants of consideration. Contextual Engineering research has demonstrated that a great deal of information about a population and its context may be gathered remotely through data analysis, but in the end, it should only be considered a supplement to the contextual understanding that is cultivated onsite (Timmons, 2021). This research found that a particular risk of relying on remote data for context identification precludes the important condition of accidental discovery, when the investigator encounters an unexpected or enlightening condition that enhances her understanding of

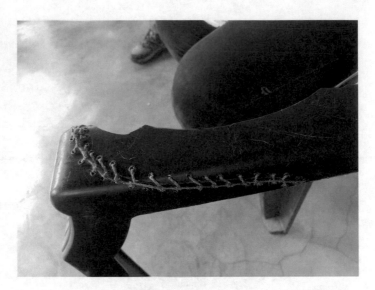

**Fig. 5.2** Repairs to a plastic chair in Sierra Leone

the population. Timmons provided the example in her research of a plastic chair at the home of a tribal leader in Sierra Leone (Fig. 5.2) that was repaired through Innovative Self-Sufficiency.

Had she relied solely on the big data she had used to assess this community's context, she would not have observed the creativity applied to extend the life of the chair, preventing her from understanding the full context of the rural population. Only by visiting the community and finding herself seated on this repaired chair did she gain contextual insight. Timmons recognized a value of collecting remote data, however, because it allows a Contextual Engineer to progress beyond the awareness stage, better aligning the investigator to achieve an assimilative perspective during direct interaction with the user population.

We acknowledge, then, that contextual insight may at least be initiated through the exploration of the four quadrants from a distance. So let's think a little more about the relevance of each quadrant. The global quadrant in particular challenges the Contextual Engineer to think beyond the moment to identify any preconditions that may have contributed to the local context. An easy example is a historic relationship between a population and the society from which the engineer originates. For me, as a white Western engineer of European descent, it is difficult to enter an indigenous territory in the southwestern U.S. without recognition that my ancestors likely participated in or sanctioned the genocide and disenfranchisement of Native Americans. I need to be aware of that horrific history and acknowledge my society's role in it—and my own advancement from it—if I am to engage in any meaningful way with a resident of the tribal territories. When I work with communities in Cameroon, I must be cognizant of the conflict between the portions of the country colonized by the French or by the British and I must recognize that the fractionalization of the nation by European association is overshadowed by the influence of imperialistic domination of all Cameroonians. These are not abstracts when working with others. They tangibly influence perceptions, biases, values, and relationships and need to be identified and acknowledged as appropriate. One way to think of the global quadrant is to consider yourself vacationing in a new place and thinking about what you should pack. Everything that's in your suitcase is selected based on your understanding of that place, and the things you unpack and wear will say a great deal about how you understand, approach, and interpret the place you're visiting. The residents of that place, meanwhile, will observe your suitcases, your clothes, and your demeanor and pass judgment on you based on their past experiences with tourists. All the experiences, attitudes, and assumptions that feed into how you pack your bags and whether your hosts approve of your luggage are a part of the global quadrant.

Next, we explore the local quadrant. This quadrant becomes more project-specific, both because it requires you to cultivate an understanding of the physical conditions that determine the technical operation of your design and it demands that you understand not just the place but the people for whom you're designing. Are they comfortable running equipment? Do they share a common identity or are they divided in culture or power? Are they comfortable with risk? Will they take a chance on something new or do they want something that's old and reliable? Do they do anything that's unique to their population, such as farming practices, cooking and cleaning, and child care? Are they ambitious or are they comfortable with what they have? These questions become more difficult to parse out from a distance, but a thorough consideration of the questions a Contextual Engineer could explore would be useful for diving into explorations of this quadrant.

The third quadrant, the people, challenges the Contextual Engineer to consider the expectations each participant in the project may have for its completion. As we discussed in Chap. 1, the definition of project success is dependent upon the objective of the participant, and a functional outcome for the user isn't always the driving motivation for the stakeholder. How well does each stakeholder reflect the context from which they emerge? Can they state their definition of a successful project honestly and self-reflectively? Might there be some power dynamics embedded in the stakeholder relationships that could cause issues as the project moves forward?

Finally, the fourth quadrant challenges the Contextual Engineer to consider the design process itself, overlaying the thought processes associated with the other three quadrants to make sure there's a clear understanding of user needs, a process by which the user perspective is heard and incorporated, an identification of any existing technical practices or knowledge that may be incorporated into the design, and consideration of whether the standards under which the designer typically works are relevant to achieve the desired outcome for the user population.

The Contextual Engineer won't come away from the Four Quadrants investigation with all the answers, or even with all the questions, but this stage of the contextual method will allow the designer to consider the conditions of people, place, and time that come together to drive a technology or infrastructure development. And it will assist the investigator in watching for conditions of global and local, people and process while onsite to better grasp the user population from an assimilative perspective.

### 5.2.3  The Five Influences

We finally reach the point of applying all our knowledge and assembling all of our observations for the purpose of determining the relative importance of each of the five contextual influences on the user population's technology or infrastructure.

This process involves going to the site, spending time engaging with the community on their terms rather than the terms of the investigator, and absorbing as much information and context as possible. The goal isn't to arrive with lists of questions, surveys, meeting agendas, and educational brochures. Instead, the contextual investigator should approach the community with the objective of learning from them about life, attitude, capability, and belief, regardless of whether the lessons learned appear relevant to design. Just as Timmons learned about Innovative Self-Sufficiency while sitting in a plastic chair in rural Sierra Leone, the Contextual Engineer's purpose is to absorb those bits of knowledge that aren't recorded in data. You learn something when you're kept awake half the night by roosters crowing and dogs barking, or by the steady traffic through the town with blaring music and flashing lights. You learn something by answering children's questions about you and asking them questions about themselves. You learn something by presenting yourself as someone who has come to learn the creativity and capability to thrive in a climate that you yourself find unfamiliar. You learn something by acknowledging that while you know how to build or make things at home, you want to learn how your user population builds and makes things that function well (or don't) so that you can become stronger in your own craft.

So how do you keep yourself on track to evaluate the contextual influences if your primary purpose on site is to be and to see? A Contextual Engineering Predictive Tool[1] initially was developed and tested as part of a doctoral dissertation exploring the significance of context in engineering design (Witmer, 2018). This tool is unique in that it does not analyze the outcome or performance of a technical design, nor does it track changes in the adoption or maintenance of infrastructure. Instead, its purpose is to provide guidance and understanding in advance of design so that the knowledge gleaned from its use

---

[1] The Contextual Engineering predictive Tool is the intellectual property of the University of Illinois Urbana-Champaign and may be accessed by creating a free account at https://contextual.engineering.illinois.edu. Processing of tool results is performed after a completed tool sheet is uploaded to the same web portal.

may be incorporated into a more robust, societally appropriate infrastructure solution. It has been adopted by international-service organizations as a diagnostic process for design teams, and variations of the tool have been created for use with domestic engineering projects. The tool consists of 41 questions for the technical designer to answer by observing during site investigation a variety of local conditions that range from the way people interact and the processes they employ in decision-making to the accessibility of education and public services in the user community. By scoring each of the questions on a Likert scale of 1–5 after completing an on-site assessment trip, Contextual Engineers are challenged to think deeply about the differences between their own experiences and the community's, pushing them to assimilate rather than simply bear witness to conditions they could easily observe without thought. The intent of the tool, then, is really to provide Contextual Engineers with a set of prompts by which they may reflect on what they observe and develop an understanding of the non-technical conditions that may affect the outcome of design decisions. In reality, the tool is a measure of the evaluators' perceptions of the user conditions rather than an objective reality, though it does provide a set of values of relativity among the five contextual influences.

Questions asked in the tool do not overtly address the influences being evaluated, and designer-users likely will be unable to influence tool outcome based on assumptions of questionable relevance. In fact, the questions examine conditions that are believed to be associated with one or more influences being investigated on a very subtle level.

The questions themselves were derived from extensive personal experience in observing and assimilating with user societies while practicing engineering design in diverse locations throughout the non-industrialized rural world. Because the literature is largely silent on identifying influences, a series of questions was developed experimentally and initially tested during travel to determine whether the outcome scoring conforms to the experience of the author. Validation of the questions is an ongoing process that will be dependent upon continuing use of the survey and feedback from users regarding their perceptions of the tool process and outcome. For this reason, the tool is distributed under a licensing agreement that requires users to allow their submitted data to be used for ongoing investigation of tool validity and functionality.

The directions given to tool users when they obtain the questionnaire state that travelers should review the questionnaire in advance but resist answering questions until they have spent significant time in the user population. For those questions with which the technical designer struggles, the guidance recommends making a best-guess after the trip has ended so that all items have been completed by the time the tool is submitted for scoring. The scores range from 1 to 5, with 1 signifying the absence of a condition and 5 signifying the prevalence of that condition. Some of the questions may be easily answered without interacting with the client, while others may require conversations with community members, attendance at meetings, or passive participation in public events. None of these questions are intended to be asked directly of community members, however. Many of the questions are deliberately vague, for the purpose of encouraging the user to think more deeply about the purpose of the question and how she might resolve any ambiguities in her own mind. Technical designers using the tool are strongly encouraged to try to answer every question, even if they must guess, rather than avoiding the topic and leaving the score blank. The author recognizes that the user's responses to the questions may be influenced by their own experiences and predispositions, though the questions are intended to circumvent some of the more obvious biases an industrialized-world user may bring to the process by observations that may not appear overtly relevant to the technical designer. Though the tool was designed to minimize the importation of predispositions or beliefs held by the user, it is important to recognize that inherent biases lie in all humans, and particular events or experiences may have an impact on a user's interpretation of the observation.

For example, by asking the technical designer to consider whether homes in the user population contain upholstered furniture, the tool prompts the technical designer to weigh user conditions against the conditions she may take for granted in her home surroundings. The question may seem irrelevant, and had been challenged by a few users as appearing trivial when they initially received and reviewed the checklist, but upon further reflection, those users said they recognized that upholstered furniture may carry multiple meanings in a different society. A sofa, for example, may cost a great deal more than a plastic chair. It may be difficult to transport and require resourcefulness on the part of the owner to purchase and move it to the home. It may be more prone to collecting dust and dirt so that it would require more cleanliness or more diligent hygiene to preserve it for many years. And it may convey a sense of pride, a connection to other experiences, or a desire to appear "advanced." The very act of considering the question should lead the user to become not only more attuned to the furnishings in homes but also to the lifestyles adopted by the client, which expands the user's understanding of her client's values and motivations. If a user comes from a background in which furniture is sparse or simple (the minimalism of traditional Japanese furnishings, for example), she may interpret both the question and its meaning differently than a user from a background that relies on reclining chairs and high-loft beds. When the Predictive Tool is processed, the calculator initially determines a user's raw score for each of the five influences. It cannot be emphasized strongly enough that the raw scores by themselves mean nothing, and the magnitude of a score is insignificant compared with its relation to the other influences being considered. Final scores are *relative* rather than absolute, indicating the magnitude of each influence out of 100%. In other words, if each influence were equivalent in significance, each would score at 20%, so it is the deviation from 20%, either higher or lower, that demonstrates the importance of the influence on technology adoption. An example of Predictive Tool results is shown in Table 5.1. To visualize the significance of relative influences more easily, one may picture a netted stress ball, which appears symmetrically round when held in the palm of one's hand. Applying pressure with a finger at a particular location, however, causes a blister in the ball to form and pop through the netting, and consistent application of the same amount of pressure at any point on the ball will theoretically result in

**Table 5.1** Sample results of Predictive Tool output for relative contextual influences as calculated for a rural community in Nicaragua

| RESULTS | | |
|---|---|---|
| Influence | Raw Score | Weighted Score |
| Cultural | 2.39 | 15.8% |
| Political | 2.96 | 19.6% |
| Educational | 3.50 | 23.1% |
| Mechanical | 3.18 | 23.8% |
| Economic | 2.68 | 17.7% |

a blister emerging at the same spot with equal consistency. Similarly, a community may appear to be in balance for all influences during normal daily life, but the application of a new infrastructure intervention by an outside technical designer can generate relational "blisters" within the society that result in failure to adopt the infrastructure, discomfort with the process, disdain for the complexity of the infrastructure, or anger with the cost associated with its construction. The use of the Contextual Engineering tool, then, is intended to identify where those blisters will emerge and prevent them from growing so large that they subvert the objective of intervention.Some general guidelines associated with each influence, shown in Table 5.2, encourage designers to think about the impact that design decisions may have upon the user society. Harking back to the community of Llano Largo, Honduras, that was discussed in Chap. 1, a stronger understanding of societal drivers may help an engineer determine whether a system utilizing three water sources would be acceptable or disastrous for the client. Using the general guidelines, one can predict that if the community demonstrates a high political influence, the three-source system could conflict strongly with the guideline to "recognize existing power structures of user population and adhere to those structures to prevent conflict." A low mechanical influence, similarly, could indicate that the community's leaders may lack the technical expertise to operate a pump even if they express a willingness to accept it; care would be needed before implementing advanced technology. Taken together, the guidelines can provide a roadmap for the technical designer to explore technical alternatives and weigh them on a sociocultural scale in addition to a technical scale to determine their appropriateness for user adoption and operation.

## 5.3 Pre-travel and Post-travel Perceptions

The accidental completion of the Predictive Tool before travel by one professional EWB technical designer provided a glimpse into how a design-team member may perceive a

**Table 5.2**  General guidelines for application of contextual influences to technical design decision-making

| If this influence… | …is high | …is low |
|---|---|---|
| Cultural | Must have strong respect for cultural constraints and beliefs, heavy and frequent input of user | May be less sensitive to identities, behaviors, or beliefs in selecting design features |
| Political | Must recognize/respect power structure and input of the powerful; be sensitive to voices that matter but are unheard. Community is likely very heterogeneous | Communal decisions may govern over those of individual power holders; less necessity to conform to power structure's conditions and expectations |
| Economic | Self-sustaining, low-maintenance designs are best; infrastructure should not demand frequent time/money investments | Autonomy in maintenance and operations is acceptable; user population may be more willing to accept projects that require continuing cost |
| Educational | New approaches that advance existing technical understanding are likely to be well received; user may welcome promoting new technology to others | Infrastructure should incorporate familiar, preferably indigenous technical features; avoid designs that require extensive training |
| Mechanical | Adapt existing technologies to new technical design to keep them familiar; engage user in design process to assure comfort with processes | Avoid exceeding user's technical experience with design; be wary of user rubber-stamping technical design out of unfamiliarity |

user population, relying on her own information filters and the veracity of her information sources, before conducting a firsthand on-site investigation. Table 5.3 shows the calculated relative influence results of the Predictive Tool responses before and after travel to the user community in Panama for this individual. Before visiting the community, she perceived that members of the user population struggled to meet their perceived basic needs (Economic Influence) and this governed their ability to adapt and operate an engineered infrastructure. But after she visited and completed a new predictive-tool questionnaire, her understanding of the community shifted strongly toward recognizing the influence of a strong local identity, values, and beliefs (Cultural Influence), which became the predominant infrastructure-acceptance determinant identified by her tool scoring. This switch in perception aligned closely with those of a fellow technical designer who traveled at the same time and completed only a post-travel questionnaire, as is shown in Table 5.4, as well as the majority of translators and field-support local resources and community liaisons who worked with the travel team.

Interviews with the technical designers after travel indicated that they realized after reviewing the tool results that they did not actually possess a full understanding of the

**Table 5.3**  Predictive tool results for a technical designer who completed pre-travel and post-travel questionnaires for Ngabe-Bugle Comarca, Panama

| | RESULTS | |
|---|---|---|
| *Influence* | **Weighted Score pre-travel** | **Weighted Score post-travel** |
| Cultural | **19.9%** | **26.1%** |
| Political | **19.5%** | **19.3%** |
| Educational | **20.3%** | **19.7%** |
| Mechanical | **17.7%** | **17.7%** |
| Economic | **22.5%** | **17.3%** |

**Table 5.4**  Comparison of scores for translators, community liaisons, local resources, and EWB traveling team members (red cell is the greatest influence for a given community, green cell is the least)

| Subject | Cultural | Political | Educational | Mechanical | Economic |
|---|---|---|---|---|---|
| TM 1 (pre) | 19.9% | 19.5% | 20.3% | 17.7% | 22.5% |
| TM 1 (post) | 26.1% | 19.3% | 19.7% | 17.7% | 17.3% |
| TM 2 | 26.6% | 20.3% | 18.6% | 12.8% | 21.6% |
| Trans 1 | 26.1% | 20.0% | 19.5% | 12.8% | 21.6% |
| Trans 2 | 21.9% | 23.3% | 17.1% | 18.0% | 19.7% |
| Trans 3 | 25.7% | 19.9% | 20.2% | 17.4% | 16.9% |
| CL 1 | 26.5% | 20.9% | 18.8% | 13.1% | 20.8% |
| LR 1 | 20.4% | 22.5% | 20.0% | 18.1% | 18.9% |
| **Collaborative**[1] | 26.2% | 21.0% | 18.0% | 14.0% | 20.7% |

KEY: TM = team member; Trans = Translator; CL = community liaison; LR = local resource.
[1]Collaborative group = TM 1, TM 2, Trans 1, Trans 2, CL 1

society with which they were working before they traveled, though they had believed they were fully cognizant of local conditions. The image they had constructed pre-travel was built mostly from the information they had received through correspondence and conversations with the community liaison, whom they deemed to be honest and forthright in communications but unable to fully convey a clear picture of conditions in an unfamiliar community to the EWB professional team. As a supplement to the information provided by the liaison, the team relied upon data gleaned from web searches and available publications, again deemed to be accurate and honest in their depiction of the Panamanian society but incomplete in their descriptions.

Because the unexpected opportunity to compare pre-travel and post-travel perceptions yielded startling differences in community understanding for the subject technical designer, an additional trial was performed in which an entire team's understanding of their own user population was tested before travel and compared to post-travel perceptions, using the Predictive Tool. In this case, six members of a university EWB team traveled to coastal Ecuador in January 2020 after spending nearly a year researching their client's conditions and identity. Team members filled out the Predictive Tool individually and then completed it collaboratively, negotiating scores for each of the 41 questions to reach a consensus. The team then spent eight days on-site, following the Contextual Engineering methodology of investigating physical, societal, political, and economic conditions, before completing the tool once more both individually and collectively. The results of pre- and post-travel relative influences identified by the team are shown in Table 5.5.

One can see from the color scale of relative influences for each individual technical designer that perspectives before travel varied widely among the group in identifying the most critical influence, with two members viewing cultural influence as most significant, two members viewing political influence as dominant, and one member each viewing educational and mechanical influences as most critical. Not one of the technical designers,

**Table 5.5** Individual and team perceptions before and after travel for El Guarango, Ecuador Project Team indicate that site observation produced strong conformity of understanding (red cell is the greatest influence for a given subject, green cell is least)

| | RESULTS | | | | |
|---|---|---|---|---|---|
| | **Pre-Travel** | | | | |
| | Cultural | Political | Educational | Mechanical | Economic |
| *Individual* | 21.9% | 19.1% | 20.0% | 18.2% | 20.8% |
| | 22.4% | 24.2% | 13.7% | 18.6% | 21.1% |
| | 18.0% | 20.7% | 22.9% | 18.6% | 19.8% |
| | 22.4% | 18.7% | 19.0% | 23.0% | 16.9% |
| | 23.2% | 27.1% | 12.6% | 12.2% | 24.8% |
| | 23.1% | 22.2% | 17.6% | 15.4% | 21.8% |
| **Group** | 21.7% | 20.2% | 18.7% | 18.1% | 21.2% |
| | **Post-Travel** | | | | |
| *Individual* | 18.5% | 18.9% | 20.5% | 19.5% | 22.6% |
| | 17.2% | 19.8% | 21.4% | 19.2% | 22.5% |
| | 19.3% | 20.2% | 20.0% | 19.1% | 21.4% |
| | 18.8% | 21.6% | 17.4% | 19.0% | 23.1% |
| | 14.3% | 20.7% | 21.4% | 19.5% | 24.1% |
| | 14.1% | 21.5% | 20.5% | 18.7% | 25.1% |
| **Group** | 16.5% | 19.0% | 20.7% | 19.1% | 24.7% |

however, perceived the economic influence as strong within the user population, even when the team negotiated their pre-travel scoring collectively. After travel, however, all six technical designers viewed the user population's contextual influences similarly, at least in terms of most and least significant relative influences. And those individual perceptions also aligned strongly with the group-negotiated tool outcome, which identified as the most dominant influence on the community's inability to meet what it considers its basic needs; the group outcome also found that the least significant influence was a commonly shared and valued sense of identity that aligns with a set of values and/or beliefs.

## 5.4 Conclusions

The process of investigating context relies upon the 3-4-5 method, which begins with self-appraisal to calibrate the investigator's perspective through three levels of perception; then considers the four quadrants of context—the global, local, people, and process—before investigating the five contextual influences that determine a user population's likelihood to adopt and maintain a technical infrastructure or process.

A Predictive Tool may be used to calculate the relative significance of each of the five contextual influences for a user population, and the tool consists of questions the investigator answers after observing and interacting with the population.

The relative significance of each contextual influence may be pondered in advance, but a deep exploration of context is best performed after the technical designer has had an opportunity to self-calibrate and interact directly with a user population.

**Questions for Parable Consideration**
1. The suggestion to implement ultraviolet disinfection technology was not only contextually inappropriate but also technically inappropriate as well. With this in mind, what do you think drove this decision?
2. What could be some unintended consequences of this failed design proposal?
3. At which point did the engineers of this project begin to break down the global, local, people, and/or process (if at all)?
4. Consider all possible motivations of the manufacturer's representative to donate the UV systems? Which would you characterize as explicit and which would you consider implicit?
5. If you were the lead technical designer on this project, what would you have said to the UV donor? The community? The other project stakeholders?

# References

Timmons, A. C. (2021). Rural-urban contextual data triangulation for international engineering project work (Master's thesis). University of Illinois Urbana-Champaign.

Witmer, A.-P. (2018). Contextual engineering assessment using an influence-identification tool. *Journal of Engineering, Design and Technology, 16*(6), 889–909. https://doi.org/10.1108/JEDT-05-2018-0091.

# Applying Context to Engineering

**The Parable of the Skeptic**

The Contextual Engineering Research Group (CERG) at the University of Illinois is a vibrant, multidisciplinary team of undergraduates and graduate students, who bring diverse backgrounds, experiences, and expertise to engineering technology but who share an interest in matching that technology with the people for whom it's intended. The group has become well enough known on campus that an average of five to ten students per semester ask to attend CERG meetings in preparation for conducting contextual research of their own. Some of these students subsequently undertake ambitious research projects; others prefer just to learn about the diversity of contextual applications. After one meeting, a relatively new undergraduate approaches the CERG leader to ask a pointed question: "How do you know Contextual Engineering works?" The answer does not come easily because the discipline is in its infancy, and designs implemented using Contextual Engineering have not had the opportunity to withstand the test of time. So the leader responds: "We don't!"

Dissatisfied, the student asks another way: "How do you justify this process if you can't prove its value?" Responds the CERG leader, "I know its value from the feedback I get from user communities, who express confidence in their infrastructure rather than gratitude for its delivery. They exhibit pride in the work they've done, not appreciation for the work of others. And they look for ways to improve the infrastructure from the day it's built, rather than fearing that their actions will anger those with whom they've worked to create it." The student looks puzzled. "Is that good enough?"

© The Author(s), under exclusive license to Springer Nature Switzerland AG 2022  85
A.-P. Witmer. *Contextual Engineering*. Synthesis Lectures on Engineering, Science, and Technology, https://doi.org/10.1007/978-3-031-07692-3_6

## 6.1   Applying Context

By now, we've explored the importance of context, how and why it differs from traditional engineering design practice, and techniques for investigating its unique characteristics in a given community. Now, we need to consider how to apply this newfound knowledge and insight to technical decision-making processes, and here's where the water gets muddy. How do you take sociocultural knowledge and apply it to technical design, particularly when the whole notion of using context to inform design is such a neoteric practice?

We start by returning to the acknowledgment that whether we know it or not, we have been trained as engineers to think deterministically about technology. We follow a technical problem-solving method that suggests there is a single correct path that leads from problem identification to the solution, and that path follows the deterministic assumption that the end result will be superior technology, regardless of whether the user population appreciates its value or not. But if we overlay our design process with the contextual understanding that we've accumulated here and accept that technology is not independent of the society that creates or uses it, we begin to recognize how often we have assumed our decisions are preordained when in fact they are highly variable.

Consider the design of a water system for a community in rural Nigeria, developed and constructed by a student organization (Fig. 6.1). Engineering teams spent countless hours visiting the community, taking measurements, talking with leaders, organizing governmental structures for water system management, and performing detailed calculations that would make a practicing U.S. engineer's head swim. The team taught the community how to build ferrocement storage tanks, which they themselves learned to construct during the process. They obtained grant monies to buy a quarter-million-dollar diesel generator to power the submersible pump, which they selected to draw supply from a deep borehole. They returned to Nigeria frequently to make repairs, and they continue more than a decade later to check in with the community many times a year to answer questions and address problems with operation.

The engineering design for the structure was sound: a borehole pump draws supply from the aquifer underlying the community to fill two storage tanks, which are then drained by gravity to public taps. The community water committee was charged with hiring tap stand operators, who collected money from residents in each of the community's six kindreds as they filled their vessels. The wealthiest members of the community were given the option of purchasing a direct-to-home service line, allowing them to have indoor plumbing for the first time. Each facet of the design was technically exquisite, with carefully calculated flow rates and pressures, peak demands and tank fill–drain cycles, construction documents for borehole and generator, and reams of guidance documents for community overseers. And yet, members of the original team, long out of college and practicing their engineering professionally for a decade, still find themselves tapping their own savings accounts to help the community with equipment repairs and fuel costs. They

**Fig. 6.1** Residents of Adu Achi, Nigeria, quickly assemble to collect water from an EWB-designed borehole during test pumping

still negotiate contracts with workmen on behalf of the village, or at least provide guidance to community leaders on how to seek quality repairs. And the oversight committee they helped to establish has been replaced by at least two newer incarnations of a water oversight board, each of which has been as ineffective as the last at keeping the system functional. One could easily attribute this never-ending infrastructure struggle to a lack of societal understanding, since the technology was exhaustively detailed in design for functionality. So let's take a look at where things went wrong with the technical process.

We'll begin with the conception of how the system was planned to work. Because of some political strife in the region during the years leading up to construction, it was decided that it would be wisest to place the borehole directly next to the *Igwe's* palace. (In Igbo Nigeria, the *Igwe* is a combination of the mayor and chief—both the jurisdictional officer and cultural leader, and the *Igwe* of this community was a wise, honest leader. But he also was subject to Nigerian practice and history, accustomed to accepting his share of authority and influence that accompanies the position of a dignitary.) This allowed the team to place the diesel generator within the walls of the palace, providing a power source not only for the well but for the compound as well. The tanks were constructed adjacent to the borehole on the second-highest land in the village so that they could serve all residents EXCEPT for one kindred. The exact reason for exclusion is not clear and may have been the result of physical elevation or social discrimination. The piping system was designed to flow by gravity from the tanks to public taps in each of the kindreds, with tap stand operators earning a wage by collecting fees from users. Only the wealthiest residents in the community could afford to tap into the pipeline to draw water directly to their homes because the system designers and water board agreed this could be a healthy income opportunity for diesel fuel and repair supplies.

Already, a contextual knowledge of the community demonstrates the design process was flawed. The exclusion of one kindred from the system created political tensions that rippled through the community and made it difficult to engage any of the residents in joining the oversight board. In response to the colonization of the nation and its long struggle to free itself from British rule, a system of subterfuge in the form of corruptive practices such as embezzlement became the norm. It was no surprise, then, that financial reserves for the system that were supposed to be collected by the tap stand operators hired to collect fees from users never accumulated. Instead, the operators reported each day that no one was using the water, pocketing the fees they had collected for themselves instead of turning them over to the water committee. The pressurized pipe design made no sense to local plumbers, who were accustomed to speeding through pipe installation without ensuring joints were tightly sealed. As a result, whenever the tanks were full, the entire system leaked like a sprinkler, often bubbling up through the ground. To keep the system pressurized so that no microbes could contaminate the water from infiltrating ground moisture would mean running the diesel generator several times a day to pump replacement water into the tanks, a costly and futile process. Once wealthy residents purchased home access to their water, they began to share that water with their families living nearby, further reducing the number of people going to the public tap stands and paying for their water. And the list goes on and on and on.

Would Contextual Engineering have prevented some of these design errors that led to a system that remains nonfunctional more often than it's usable? I believe so. Table 6.1 shows the Contextual Fingerprint as determined by the Predictive Tool completion of four project participants, whose scores were averaged together after they completed their

**Table 6.1** Individual results of the Predictive Tool for four project participants, as well as an average of the individual results

| Nigeria Group | | | | | |
|---|---|---|---|---|---|
| INFLUENCE | #1 | #2 | #3 | #4 | **Avg** |
| Cultural | 19.7% | 20.5% | 19.9% | 19.9% | **20.0%** |
| Political | 19.9% | 20.6% | 20.5% | 21.6% | **20.6%** |
| Educational | 21.4% | 22.3% | 22.1% | 22.1% | **22.0%** |
| Mechanical | 20.4% | 16.8% | 16.6% | 16.6% | **17.6%** |
| Economic | 18.7% | 19.8% | 20.9% | 19.9% | **19.8%** |

individual observations on three separate trips. As you can see, while the magnitude of relative influence differs a bit from person to person, individuals completely agreed on the largest influence, and they align with each other on the bottom-two and top-two.

Based on the contextual fingerprint, this community's most important influences were educational and political, while the least important influence was mechanical. What does that mean? The general guidelines shown in Chap. 5, when examined for the Nigerian community, indicate that residents were curious and liked to learn new things, but they were also very heterogeneous in opinion and structured around power and influence. Also, they didn't care too terribly about keeping things running or investing a lot of effort in sustaining an operation.

Had the original design team been able to perform a contextual analysis of this sort, the results would have provided a big red flag regarding the exclusion of one kindred among the six from direct access to water. This political exacerbation of already strained relationships could have warned the team that a collaborative water board drawing from all six kindreds, when only five have direct access to water, could lead to organizational failure, not to mention hostilities (which did occur!). New technologies wouldn't be a deterrent to community adoption, but those new technologies had better not require too much oversight because the residents prefer not to be bothered with daily operation and maintenance tasks (low mechanical influence). The cost of water wasn't an overwhelming concern for residents because at least a good portion of the community feels it has the resources it requires to meet its own needs, which was the case as more and more residents opted to purchase home connections and forego using the tap stands. Pressurization of the system, however, required constant operation and maintenance to keep the generator fueled and running, which all but assured that the system would drain regularly in this laissez-faire society.

These are some of the direct findings from the tool, but the investigation of the contextual conditions can provide additional design and construction clues. A global examination of Nigeria as a society demonstrates that colonization and exploitation of the indigenous population were prevalent for more than a century. The colonial influence has been recognized as a prime catalyst of transactional attitudes toward relationships and comfort with bribery as a way of doing business (Kroeze et al., 2021), and Nigeria is a prime example of the corruptive influence of colonialism. People in this rural community participate in a project, but only to the degree that they are compensated for their effort. They are warm and generous, but they also are pragmatic when providing a service or effort. And they are as comfortable with purchasing another's capabilities as they are with asking for compensation for their own. After completing detailed borehole design documents, the student group charged the community with hiring a contractor to drill the well. A driller from the community, who had a superior reputation for workmanship, offered his services to the community in the form of bribes to the water board members. Another driller from outside the community, with a less stellar reputation and a higher fee, also submitted for the work but offered no bribes. The students insisted that the water board hire the outside driller because he didn't violate the student designers' own moral objections to bribery.

It comes as no surprise after that incident that the community has never referred to the water system as its own, always referring to it instead as belonging to the student organization. Within a decade, the borehole collapsed and required reconstruction, the cost of which was borne by the now-former students who raised funds and raided their savings accounts to pay for the work.

The list of design and construction errors—including *technical* errors—that occurred despite the unparalleled effort by the student engineers is lengthy. But it's important not to fault the students for the ongoing system woes in this small Nigerian village, because they had followed the conventional engineering process for rigorous design and societal exploration while keeping each of them separate and distinct from each other, as we've all been taught to do.

So how can we use Contextual Engineering when performing engineering design on our own projects? Of course, the first step is to complete the 3–4–5 Method to build a solid if not an entirely complete understanding of the conditions of the user community. This means identifying the contextual influences that govern and the influences that are insignificant, but it also means building intuition about what really matters to the population and what might be a make-or-break feature of technical design.

Once that understanding is in place, it's truly time to design, exploring all the options— not satisficing by selecting the first reasonable design option to present itself—and evaluating them based on what you know. We often use a Decision Matrix in our Contextual Engineering courses to consider the strengths and weaknesses of each design option, weighting them by looking at the contextual fingerprint and scoring them as a team to augment each team member's perspective with the observations of the others. This is the very same process many professional engineers use in the U.S. while working to identify an optimal approach to technical design (minus the fingerprint weighting). Table 6.2 shows an example Contextual Decision Matrix that was constructed to demonstrate how community contextual fingerprints may be used to determine the types of FEW (Food–Energy–Water nexus) technologies best suited to an Andean user population's needs.

The leftmost column of this Decision Matrix lists off approach alternatives in the areas of water (blue), energy (red), and food (green) that were proposed for application to a series of indigenous Aymara communities located high in the Andes Mountains. The approach alternatives were defined by a team of engineers working in partnership with a local NGO to explore more than a dozen villages of differing sizes, locations, and characters. All approaches identified by the team, drawing not only on their technical expertise but on the expressions of interest or capabilities they noted while interacting contextually with the villages, were included in the list regardless of cost, viability, or applicability.

The team then collaboratively completed the Predictive Tool for each community. The resulting relative influences determined by the tool were placed in the "weighting" row,

**Table 6.2** An example decision matrix using community Contextual Fingerprint for one village in the Aymara region of the Bolivian Andes

| Scoring of Technical Options | | | | | | |
|---|---|---|---|---|---|---|
| Approach | Implementation Cost (Econ.) | Comfort with Use (Mech.) | Breadth of Application (Pol.) | Design Comfort (Cult.) | Adaptability Evolvability (Educ.) | SCORE |
| *weighting* | *0.18* | *0.18* | *0.22* | *0.25* | *0.17* | *1* |
| Solar Pumping Equipment | 3 | 2 | 3 | 3 | 3 | 2.82 |
| Well Development | 4 | 5 | 5 | 5 | 2 | 4.31 |
| Chlorine Treatment | 4 | 3 | 5 | 4 | 2 | 3.7 |
| Source Water Protection | 5 | 4 | 5 | 3 | 4 | 4.15 |
| Fog Catcher Technology | 3 | 3 | 3 | 2 | 5 | 3.09 |
| Solar Panels for Home Power | 2 | 3 | 3 | 2 | 3 | 2.57 |
| Conversion to LED lighting | 3 | 5 | 4 | 4 | 2 | 3.66 |
| Solar Food Cooker | 2 | 3 | 3 | 3 | 3 | 2.82 |
| Drip Irrigation | 1 | 4 | 4 | 4 | 5 | 3.63 |

and a different table was created for each community (Table 6.2, then, represents only one community among the dozen evaluated). Finally, team members iteratively and collaboratively assigned scores to each technology—independent of community context—on a scale of 1–5 for each of the approaches. For example, a technology associated with solar panels for home power would be costly and scores only a 3 under implementation cost, while source water protection would produce no real implementation cost and thus scores a 5 out of 5. It is only when the score is weighted using the contextual influences that the value of a particular technical approach is interlinked with societal conditions to identify a most likely desirable design for a particular population. While the example presented here employed technologies predefined by project parameters (set by the funding source), the same methodology may be used for virtually any technical design exploration. The key to the process is a collaborative and iterative identification of approach, contextual condition (aligned with the influences to provide column headings), and technology scoring.

As with everything contextual, we acknowledge that completing such an alternatives analysis still may not assure that a design team has selected the contextually optimal technology for a user population. In fact, this is the first step in contextual technical design, since the designer must then present one or several of the best-scoring options for discussion with the user to gauge response, assess applicability, and further refine weighting and scoring. Only after this process has been completed iteratively with the design team and user can the Contextual Engineer feel some level of comfort in moving forward with a final technical design approach.

## 6.2    The Reality of Contextual Engineering Practice

Contextual Engineering has generated methods to quantify the relative significance of five key influences that reside within a user population, but does it create an accurate enough picture to rely upon for complex decision-making? The most accurate answer is "We don't know." But this neglects to recognize one of the most basic principles of Contextual Engineering, which is that human dynamics in interaction with technology is inherently uncertain, and the more confident we become that we know the intricacies of these dynamics, the less competent we are in acknowledging and compensating for the uncertainties. This is, to some degree, a twist on Heisenberg's Uncertainty Principle and Schrodinger's Cat, which in the parlance of early twentieth-century salon conversations were reduced to the assertions that observation is not reliable yet only observation can confirm a state of being. This paradox leads the Contextual Engineer, particularly one who has spent considerable time exploring the fragilities and limitations of their own perceptions, to conclude that even the most rigorously quantified evaluation of user-community context is flawed and may not even capture the conditions on-site.

So why do it? Let's consider the classic Senegal well story, which I've personally encountered while traveling in rural West Africa. The story (and this is a true story, related to me by the person who designed and constructed a well for a rural Senegalese community while serving in the Peace Corps) is that a well-meaning Westerner observes the burden upon Senegalese women of collecting water, which requires walks of many kilometers to haul heavy jugs of water home for consumption, cooking, and cleaning. He consults Peace Corps manuals and talks with water engineers before designing a hand-dug well in the center of town, meters away from most of the women's homes. This will ease their burden, he thinks as he rallies the men of the village to help dig the well, and allow women to be more economically productive for their families. The village celebrates the well's construction with ceremonies and proclamations, led by the village men who touts the value of this new infrastructure that would free their wives to care for children and work in the fields. When the Peace Corps volunteer returns to the town a month later, though, the rope pulley has been destroyed and the well fouled. He is surprised, but when he learns the women themselves were to blame, he is utterly stunned. He knew this community. He had lived among the rural Senegalese for many months, sat beneath the shade trees in the center of the village on woven rugs with the men and talked about the conditions that prevent residents from profiting from their work, walked with the women to collect water, dodged the daily stampede of the livestock as they were driven to the fields to graze. But he had failed to understand the loneliness many women felt as they were sequestered behind the walls of their husbands' compounds, sharing companionship only with rival wives against whom they competed for their husband's attention. Their friends were hidden behind other walls, married to different men, and the only time they saw each other was when they were permitted to leave the compound to gather water. The daily burden of carrying water for hours was unimaginable to the Peace Corps volunteer, who grew up with faucets in his home and never considered water access a demand on his time rather than an entitlement. And the men of the village with whom he interacted were wholly in favor of shortening their wives' water-fetching responsibilities so that they could perform other tasks, often in support of maintaining a leisurely lifestyle for the men who remained under the shade trees in conversation all day long.

Let's think about what ran through the Peace Corps volunteer's mind while implementing the well for the Senegalese women. He'd witnessed the inequity between men lolling in the shade and women walking kilometers to haul heavy water. He'd heard conversations regarding how much more women could accomplish to contribute to the wealth of their families if they could be freed from water hauling to farm peanuts, care for livestock, or teach their children. He'd experienced the exhaustion and tedium of walking hours at a time for water, which conflicted with his understanding of water as an instantaneous resource. But the village well had violated conditions that he hadn't come to recognize because he'd framed the problem from his own experience. Did he talk with women—outside the hearing of their husbands—about the well? Did he understand the competition that exists in polygamous Senegalese families among the wives, who often are cloistered with their rivals and sequestered from their friends?

Without challenging himself to question the unknown unknowns, he would not question the value of a well in the community. Perhaps if he had quietly observed the women he walked with as they gathered water rather than conversing with them, he might have better understood the feelings of the water bearers. If he had considered that a life-long daily responsibility of carrying water is not as onerous to someone who knows no other option, he may have shifted his efforts from creating a well to creating a water-carrying system. If he had traveled to other villages and observed how other men and women addressed water needs and the level of satisfaction they felt with the solutions he found, he might have shifted his focus entirely to an alternative project to support Senegalese women. Perhaps if he had challenged his own belief that an individual can be more self-actualized if given sufficient time and money to engage in non-routine activities—he may have aligned better with the conditions and needs of the people with whom he was working.

These are all purely speculation because, in keeping with Schrodinger's Cat, we cannot know whether an outcome is correct unless we observe that outcome. This is a particularly unsatisfying aspect of Contextual Engineering. We never know for certain whether we've observed enough, learned enough, or challenged ourselves enough, until we witness an outcome, learn from it, and move forward.

Engineers do not like to think of their process as one of learning by doing. We rely upon mathematical equations and scientific principles to perform our designs, and we like to consider the results as successful as long as we "do the math" correctly. But history is littered with engineering failures that resulted not from a failure of math and science but from an inability to predict the unpredictable. Sometimes the unpredictable is a force of nature, more often the unpredictable is a force of humanity. And this is where Contextual Engineering supplements traditional engineering approaches, not by predicting the unpredictable but by exploring the potential for unpredictability and accounting for the likelihood that our designs may follow a different use path than we had originally intended.

A note on my desktop monitor reminds me that engineering design for a user population should not be based upon our intent or desire for a user population but upon the likely conditions under which the user population will put it to use, maintain it, or abandon it, and evolve it into a stronger system as the population itself evolves. To do this, we must be thoughtful in questioning our own understanding and methods. Always.

## 6.3    The Contextual Engineering Paradox

We're stepping into a paradox as we talk about the need to immerse and understand the specific context that resides within a user population. Contextual Engineering is a process that attempts to generalize a technical designer's ability to identify and address specific and often exclusive conditions that reside in a place and time. But how can a process that bores in on the uniqueness of a set of conditions be generalizable to any people and place in the world?

The answer lies in the recognition by the Contextual Engineer that the inquiry process is an exploration, not a solution. Contextual Engineering equips the technical designer to consider the existence of the unknown and incorporate those unknowns into the design, which is a generalizable skill. But any Contextual Engineer will tell you that it's a recipe for failure to draw insights from one experience and apply them to another, even if there appears to be absolute parity between conditions and populations.

Let's look at an example. Contextual Engineering currently is being used to assist rural, farm-based communities in identifying solutions for a variety of conditions that plague rural populations, including the impacts upon the agriculture of global climate change and the intensification of chronic mental health disorders such as depression and anxiety. These studies do not seek to impose actions or initiatives upon the communities but to assist communities in contextually addressing responses to these amplifying conditions as they occur. Application of Contextual Engineering investigations is particularly challenging in the rural U.S. because farm communities tend to insulate themselves by distance and connectivity from the diversity of thought and experience in cities, thus preserving local identity, values, and beliefs. In fact, Contextual Engineering researchers approaching rural communities are more likely to be met with suspicion and distrust than warmth and hospitality. And yet, as researchers plumbed the community context of several rural farming towns in central Illinois, they quickly were able to ascertain that one cannot assume all farming towns share the same conditions, values, beliefs, or capabilities. Two communities, in particular, stand out in stark contrast to each other, though they lie barely 90 miles apart on the eastern Illinois prairie. The first is a predominately Caucasian, economically modest farm town of about 2,000 residents whose nearest population center is an academic, internationally diverse city. The second is a predominately African-American, economically impoverished town of about 500 residents whose nearest population center is a blue-collar, racially diverse city. Faith is very important to both communities, with nine churches listed in the first town's directory and seven shown for the second. Corn farming is prevalent in the first community, while the second is better known for horse farming and vegetable production. The first community tends to be conservative politically and boasts its status as a Second Amendment Sanctuary, while the second community tends to be more liberal politically. Should we expect farmers in the first town to react to changing conditions associated with global climate change in the same way as farmers in the second town? Not any more likely than we could expect the first town's economy to adjust to changes in farm agriculture compared with the second town's economy. It's a premise of Contextual Engineering that one can't extrapolate knowledge from one place to understand another. But a Contextual Engineer *can* apply a process for learning a community's context to other locations, international or domestic, to build a better technical solution that targets the specific need of a place, people, and time.

## 6.4    Contextual Innovation and Practice

We've been talking throughout this book about technical design, technology, and its relation to people and place. Given that we call the discipline Contextual Engineering, one may rightfully conclude that this is only relevant to technical decision-making. But our research group's diverse explorations and invitations to colleagues to apply context in their own studies have surprised us by demonstrating this: if the context is determined by collectives of people in a particular place at a particular time, then the Contextual Engineering framework is equally applicable to technical design, organizational structures, government units, or any other process that relies on determining the needs and capabilities of a group of humans. In the end, Contextual Engineering is a foundation for what we more broadly call Contextual Innovation and Practice (CIP), which uses contextual principles to understand how entities work together, problem-solve, develop, and interact. CIP helped one team of researchers identify the key conditions that led to a national student organization losing its way in attracting broad participation, and it's helping another explore how communities on the Great Lakes would prefer to stabilize their shorelines when changes in weather patterns and extreme conditions threaten to erode valuable natural lands. It's taking a hard look at how the performance of a variety of agencies—government, non-profit, grass-roots—interact to provide services to indigenous

U.S. territories in an effort to better understand which organizational dynamics function to solve problems and which ones create problems. And it's considering the relationship between the advancement of commercial charging stations and the adoption of electric vehicles to assess whether historic energy-cost constraints combine with user context to encourage or suppress clean-energy vehicle growth.

Perhaps my favorite exploration associated with CIP, though, is a novel look at the relationship between Western classical music and indigenous music, and considering it as an analogy to the way we approach engineering design with indigenous populations from an elite Western perspective. Engaging with some renowned performers of musical instruments that are exclusively used in European classical music, we're trying to learn how these instruments might be re-envisioned when engaging with indigenous music around the world. At the same time, we'll look for comparable relationships between Western and indigenous technologies, with the hope that we'll be able to contextually strip away the global weight of Western exclusivity to benefit both music and technology through the rich indigenous knowledges that hide among the world's rural populations.

## 6.5    Conclusions

The application of context is inexact but requires the engineering designer to acknowledge that contextual conditions can guide technical decision-making.

Guidelines give insight into the significance of the relative contextual influences when designing a technical infrastructure or technology, but they are not the final arbiter of whether a design is appropriate for a user population.

The use of a Decision Matrix that evaluates design alternatives, assembled following contextual investigation, will allow technical designers to hone in on technical options that are most and least appropriate for a user population. It still is critical, however, to review design features directly with the user population for confirmation of the approach before completing the design.

A paradox exists in Contextual Engineering which maintains that no technical design or approach is generalizable for all user populations, but the Contextual Engineering methodology itself is generalizable and provides a framework for addressing community needs around the world by Western designers.

Contextual Engineering has spawned a broader area of application, known as Contextual Innovation and Practice, which applies contextual thinking and methodology to domestic engineering, policy-making, organizational structures, and other processes that engage collectives who seek solutions within a particular context of people, place, and time.

**Questions for Parable Consideration**
1. How do you measure success when working on an engineering design?
2. What is the biggest challenge in exploring this new way of thinking? What do you, like the student in the parable, grapple with?
3. How important is it to recognize the existence of unknown unknowns, even if you're unable to identify their relevance to your design project?
4. Is there ever a point when a Contextual Engineer can feel fully confident in their knowledge, application of technology, and understanding of user population?
5. What about Contextual Engineering makes the most sense to you? What makes the least sense?

# Reference

Kroeze, R., Dalmau, P., & Monier, F. (Eds.). (2021). *Corruption, empire and colonialism in the modern Era: A global perspective.* Springer Nature.

Printed in the United States
by Baker & Taylor Publisher Services